Introduction to

BIOTECHNOLOGY

AN AGRICULTURAL REVOLUTION

SECOND EDITION

Delmar Cengage Learning
is proud to support
FFA activities

Introduction to
BIOTECHNOLOGY
AN AGRICULTURAL REVOLUTION
SECOND EDITION

RAY V. HERREN

DELMAR
CENGAGE Learning·

Australia • Brazil • Japan • Korea • Mexico • Singapore • Spain • United Kingdom • United States

Introduction to Biotechnology: An Agricultural Revolution, 2E
Ray V. Herren

Vice President, Editorial: Dave Garza

Director of Learning Solutions: Matthew Kane

Senior Acquisitions Editor: Sherry Dickinson

Managing Editor: Marah Bellegarde

Senior Product Manager: Christina Gifford

Editorial Assistant: Scott Royael

Vice President, Marketing: Jennifer Baker

Marketing Director: Debbie Yarnell

Marketing Coordinator: Erin DeAngelo

Senior Production Director: Wendy Troeger

Production Manager: Mark Bernard

Senior Content Project Manager: Katie McGuire

Senior Art Director: David Arsenault

Cover and Interior Design Credits:

Cow: Courtesy of ARS/USDA #K7612-17

Tomato: Getty Images

Yellow Corn: © PaulPaladin/www.Shutterstock.com

Row of Corn: © Paul Clarke/www.Shutterstock.com

Green Dots Ecological Background: © Milushkina Anastasiya/www.Shutterstock.com

For product information and technology assistance, contact us at
Cengage Learning Customer & Sales Support, 1-800-354-9706

For permission to use material from this text or product,
submit all requests online at **www.cengage.com/permissions**
Further permissions questions can be emailed to
permissionrequest@cengage.com

Library of Congress Control Number: 2011941347

ISBN-13: 978-1-4354-9837-2

ISBN-10: 1-4354-9837-2

Delmar
5 Maxwell Drive
Clifton Park, NY 12065-2919
USA

Cengage Learning is a leading provider of customized learning solutions with office locations around the globe, including Singapore, the United Kingdom, Australia, Mexico, Brazil, and Japan. Locate your local office at: **international.cengage.com/region**

Cengage Learning products are represented in Canada by Nelson Education, Ltd.

To learn more about Delmar, visit **www.cengage.com/delmar**

Purchase any of our products at your local college store or at our preferred online store **www.cengagebrain.com**

Notice to the Reader

Publisher does not warrant or guarantee any of the products described herein or perform any independent analysis in connection with any of the product information contained herein. Publisher does not assume, and expressly disclaims, any obligation to obtain and include information other than that provided to it by the manufacturer. The reader is expressly warned to consider and adopt all safety precautions that might be indicated by the activities described herein and to avoid all potential hazards. By following the instructions contained herein, the reader willingly assumes all risks in connection with such instructions. The publisher makes no representations or warranties of any kind, including but not limited to, the warranties of fitness for particular purpose or merchantability, nor are any such representations implied with respect to the material set forth herein, and the publisher takes no responsibility with respect to such material. The publisher shall not be liable for any special, consequential, or exemplary damages resulting, in whole or part, from the readers' use of, or reliance upon, this material.

Printed in the United States of America
1 2 3 4 5 6 7 15 14 13 12 11

DEDICATION

This book is dedicated to my grandfather, the late Frank Herren. Although he had limited formal education, he was a true intellect. He would have been enthralled with the advances in biotechnology and would have studied every chapter in this book. One of my big hopes in life is that I can have as positive an impact on the lives of my grandchildren as he had on me.

Ray V. Herren

CONTRIBUTORS

Barry Pate, PhD
Assistant Professor, Animal Science
College of Southern Idaho
Twin Falls, Idaho

Denneal Jamison-McClurg, PhD
Associate Director, Biotechnology Program
University of California, Davis
Davis, California

CONTENTS

Historically, a large part of the American economy has been based on our gigantic agricultural industry. The production of food and fiber is basic to the survival of human beings, and American agriculture has been quite efficient in supplying the needs of the nation and a large part of the world. The advancements made in our ability to increase agricultural production have come about through the application of scientific research. This research has led to several fundamental changes in the way we produce agricultural products. In fact, these changes have been so profound that they have been referred to as agricultural revolutions.

Mechanization, refrigeration, transportation, the development of hybrid varieties, selective breeding, and the invention of herbicides and improved insecticides have all been termed as revolutionary in improving production techniques. We are now in the midst of perhaps the greatest of all revolutions—that of biotechnology. Scientists are now beginning to truly understand the intricate workings of the cell and how characteristics are passed on from generation to generation. Of all the revolutions, biotechnology is by far the most rapidly changing science. Producers have used biotechnology for many years, resulting in many major improvements in the production and processing of food. Exciting new technologies, such as cloning and genetic engineering, carry almost unlimited potential in regard to revolutionizing agriculture.

The second edition of *Introduction to Biotechnology: An Agricultural Revolution* builds on the concepts of the first edition. Since the publishing of the first edition, many new and exciting discoveries have been made, and these discoveries have been put to practical use. Almost all the chapters have been significantly updated with new information, and all the statistics and data have been revised to reflect the latest information. The second edition is intended as a text for use in high school agriscience courses. The aim is to provide students with a basic understanding of the concepts behind the biotechnology revolution in agriculture. Research has shown that context-based instruction leads to a better understanding of basic concepts and better scores on skills tests. Topics such as cell functions, genetics, genetic engineering, the uses of biotechnology, and careers in biotech are covered within the chapters. Of course, no study of biotechnology would be complete without a thorough examination of the controversy and concerns over the use of genetic engineering, genetically modified organisms, and cloning and their potential dangers to humans and the environment. All these topics are changing rapidly, and the second edition reflects the latest changes.

Each chapter begins with clearly stated learner-oriented objectives, followed by terms that are important to a thorough understanding to the text. Each chapter closes with Student Learning Activities intended to extend learning beyond the text material. Also, there are fill-in-the-blank, true-or-false, and discussion questions to help evaluate the student's grasp of the concepts.

The complete package of teaching materials should provide the necessary tools to teach an interesting and challenging course in biotechnology as part of a program of agriscience.

Extension Teaching/Learning Materials

Instructor's Guide to Text The Instructor's Guide provides answers to the end-of-chapter questions and also additional materials to assist the instructor in the preparation of lesson plans:

Lab Manual The accompanying Lab Manual was specifically designed to give students a means of performing hands-on tasks associated with the content found in each chapter.

Lab Manual Instructor's Guide The Lab Manual Instructors Guide provides answers to lab exercises and additional guidance for the instructor.

Lab Manual CD-ROM Our lab manual and lab manual instructor's guide are both available on cd. Instructors have the capability of printing off as many copies of the exercises as needed, for as long as they own the book.

Classmaster CD-ROM The Classmaster technology supplement provides the instructor with valuable resources to simplify the planning and implementation of the instructional program. It includes instructor slides in PowerPoint, a computerized test bank, an Image Library, and Student worksheets.

Chapters begin with clear educational **objectives** for students to learn, as well as **key terms** for students to focus on while reading.

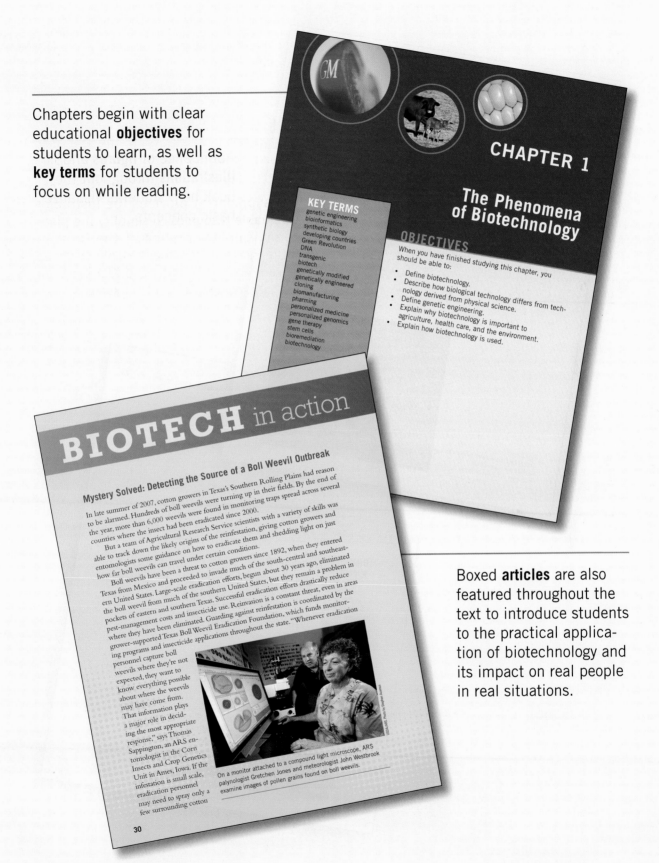

CHAPTER 1

The Phenomena of Biotechnology

KEY TERMS
genetic engineering
bioinformatics
synthetic biology
developing countries
Green Revolution
DNA
transgenic
biotech
genetically modified
genetically engineered
cloning
biomanufacturing
pharming
personalized medicine
personalized genomics
gene therapy
stem cells
bioremediation
biotechnology

OBJECTIVES

When you have finished studying this chapter, you should be able to:

• Define biotechnology.
• Describe how biological technology differs from technology derived from physical science.
• Define genetic engineering.
• Explain why biotechnology is important to agriculture, health care, and the environment.
• Explain how biotechnology is used.

BIOTECH in action

Mystery Solved: Detecting the Source of a Boll Weevil Outbreak

In late summer of 2007, cotton growers in Texas's Southern Rolling Plains had reason to be alarmed. Hundreds of boll weevils were turning up in their fields. By the end of the year, more than 6,000 weevils were found in monitoring traps spread across several counties where the insect had been eradicated since 2000.

But a team of Agricultural Research Service scientists with a variety of skills was able to track down the likely origins of the reinfestation, giving cotton growers and entomologists some guidance on how to eradicate them and shedding light on just how far boll weevils can travel under certain conditions.

Boll weevils have been a threat to cotton growers since 1892, when they entered Texas from Mexico and proceeded to invade much of the south-central and southeastern United States. Large-scale eradication efforts, begun about 30 years ago, eliminated the boll weevil from much of the southern United States, but they remain a problem in pockets of eastern and southern Texas. Successful eradication efforts drastically reduce pest-management costs and insecticide use. Reinvasion is a constant threat, even in areas where they have been eliminated. Guarding against reinfestation is coordinated by the grower-supported Texas Boll Weevil Eradication Foundation, which funds monitoring programs and insecticide applications throughout the state. "Whenever eradication personnel capture boll weevils where they're not expected, they want to know everything possible about where the weevils may have come from. That information plays a major role in deciding the most appropriate response," says Thomas Sappington, an ARS entomologist in the Corn Insects and Crop Genetics Unit in Ames, Iowa. If the infestation is small scale, eradication personnel may need to spray only a few surrounding cotton

On a monitor attached to a compound light microscope, ARS palynologist Gretchen Jones and meteorologist John Westbrook examine images of pollen grains found on boll weevils.

30

Boxed **articles** are also featured throughout the text to introduce students to the practical application of biotechnology and its impact on real people in real situations.

2 • Chapter 1

Introduction

We live in a wondrous world of inventions created by humans to make our lives better. We have all sorts of machines, devices, gadgets, and electronics that make us more efficient in our work, faster in our travel, and proficient at leisure activities. These inventions came about as a result of using the materials found in nature and through an understanding of the laws of physics. This has allowed us to fly several times faster than the speed of sound, record any type of music, develop and explore the Internet, and use the power of the atom to provide enough electrical power to supply the needs of cities with huge populations. Using the laws of nature, particularly those of physical science, to create devices is called *technology* (Figure 1–1).

Technology probably began when humans picked up stones and sticks for use as crude implements to dig in the earth or to use as weapons. Later they learned that the sticks and stones could be altered and shaped to provide a sharper edge or a better handle. As they became proficient at using these primitive tools, humans began to devise ways of making the tools better (Figure 1–2). At first, they simply used trial and error as a means of improving their inventions, but

Figure 1–1
Technology has allowed humans to develop many wondrous machines to make life better.

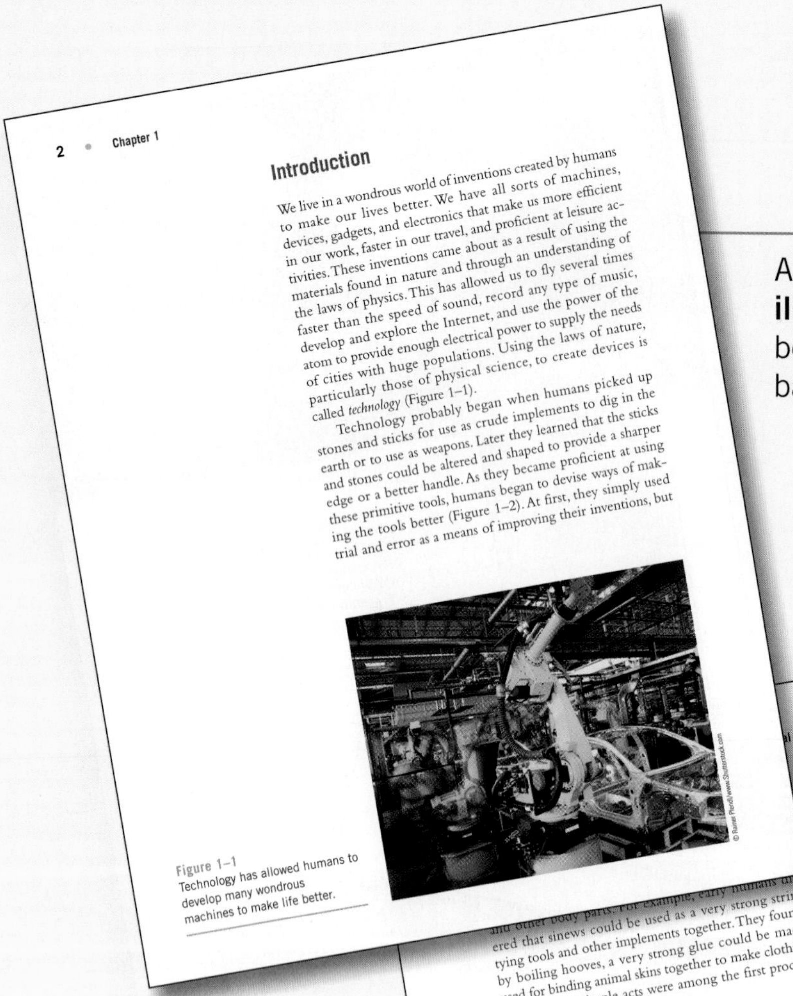

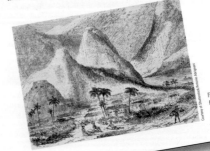

© Rainer Plendl/www.Shutterstock.com

Abundance of **figures** and **illustrations** throughout the book help students visualize basic concepts.

...l Development of Biotechnology • 23

...ans ...pt- ...ere ...of ...ns, ...and other body parts. For example, early humans discovered that sinews could be used as a very strong string for tying tools and other implements together. They found that by boiling hooves, a very strong glue could be made and used for binding animal skins together to make clothing and shelter. These simple acts were among the first processes of biotechnology.

Biotechnology and the Expansion of Civilization

Once humans could produce enough food to live in one area for an extended period of time, they had more time to develop labor-saving devices, and less people were required to produce food for survival. Individuals not involved in farming could then concentrate on developing new technologies, gathering and recording knowledge, and producing specialty goods. Early agriculture supported the development of villages and cities—civilization itself (Figure 2–2).

Courtesy of Shutterstock/Antonio Abrignani

Figure 2–2
Early agriculture supported the development of villages.

CHAPTER REVIEW

Student Learning Activities

1. Determine what you consider to be the most important milestone in the development of biotechnology. Compile a list of reasons why you consider the development to be the most important and share your reasons with the class.

2. Determine what you consider to be the most important biotechnology discovery of the future. List why you feel this technology is needed and explain what problems it could solve. Share your thoughts with the class. Try to predict when the discovery or technology will be developed and keep a list of all the class members' predictions for future reference.

3. Interview your grandparents or other elderly persons in the community. Ask them to share their ideas on what they consider to be the most important advancement in biology in their lifetime. Report to the class.

Fill in the Blanks

1. The _____ of the very best _____ and the saving of _____ from those plants was the genesis of _____ and crop production.

2. We now know that an _____ in the _____ of the calf called _____ started the process of _____ that formed cheese.

3. The early humans noticed that many _____ are good to eat and that some grasses, such as _____, produce an abundance of seed that can be ground into a _____ called _____.

4. In the mid-_____s, an Austrian monk named _____ developed a theory of _____, or how characteristics or _____ are transmitted from parents to _____.

5. Pasteur discovered that by _____ the _____ from sheep that had contracted and _____ the deadly disease of _____ into _____ sheep, the disease could be _____ in the healthy animals.

6. Dreaded _____ such as _____ were prevented by an _____ of _____.

7. They discovered that _____ is composed of _____ units of _____ and _____ and is formed in the shape of a _____.

8. Embryo transfer allowed the use of _____ females to produce _____ offspring, which, when combined with _____, allowed very _____ progress in the production of superior animals.

9. In the 1980s scientists discovered how to transfer bits of _____ information from one organism to another, allowing the _____ of _____ traits in the recipient _____ in a process now known as _____.

10. Our _____, _____, and _____ have all been greatly improved through the developments of _____.

_____ new phenomenon.

_____ uted to early bartering and money exchange systems.

_____ the first methods for storing food was by producing fruit juice and then fermenting the fruit juice into wine.

4. A purebred strain is a type of plant or animal that results from the crossing or mating of parents that are different.

5. Gregor Mendel observed his pea plants and found that short-stemmed plants produce only more short-stemmed plants and that long-stemmed plants produce only more long-stemmed plants.

6. Alexander Fleming discovered the antibiotic penicillin accidentally when his experiments with bacteria became contaminated with mold.

7. Bull semen has been successfully stored for as long as 30 years.

8. Microbes can function as natural genetic engineers and can complete horizontal gene transfer.

9. The first plant was grown in vitro in the 1930s.

10. Genetic engineering is the only type of biotechnology that scientists are currently using and studying.

Discussion

1. Explain how agriculture supported the development of cities and villages during early human civilization.

2. Describe Mendel's law of independent assortment and Mendel's law of segregation.

3. List at least three problems of human civilization that have been solved by biotechnology.

4. Who discovered the cell? What technological advance made this discovery possible?

5. Why are antibiotics considered "miracle drugs"?

6. What do you think it means to say that smallpox has been "eradicated"?

7. What are the benefits of artificial insemination and embryo transfer?

8. What is vertical gene transfer?

9. What did Watson and Crick discover? Why was this discovery important?

10. Describe the process of genetic engineering.

Chapters conclude with a thorough yet concise **review** including specific student learning activities as well as review questions to test student comprehension.

ACKNOWLEDGMENTS

The author wishes to thank the following for their contributions: Dr. Frank Flanders for his help and advice; Christina Gifford for her patience and assistance in all phases of the project; and, most of all, my wife Mary for her unwavering support and encouragement.

The author and Cengage Learning also wish to sincerely acknowledge the following reviewers for their invaluable input and for sharing their content expertise:

Stacie McKee
Whiteoak High School (OH)

Holly Amerman
R. Hungerford Preparatory High School (FL)

Thomas Vrabel
Trumbull High School (CT)

Jim Statterfiled
Jefferson County High School (TN)

Kacia Cain
Central Campus High School (IA)

Tyrone Casteel
Woodland High School (GA)

CHAPTER 1

The Phenomena of Biotechnology

KEY TERMS

genetic engineering
bioinformatics
synthetic biology
developing countries
Green Revolution
DNA
transgenic
biotech
genetically modified
genetically engineered
cloning
biomanufacturing
pharming
personalized medicine
personalized genomics
gene therapy
stem cells
bioremediation
biotechnology

OBJECTIVES

When you have finished studying this chapter, you should be able to:

*   Define biotechnology.
*   Describe how biological technology differs from technology derived from physical science.
*   Define genetic engineering.
*   Explain why biotechnology is important to agriculture, health care, and the environment.
*   Explain how biotechnology is used.

Introduction

We live in a wondrous world of inventions created by humans to make our lives better. We have all sorts of machines, devices, gadgets, and electronics that make us more efficient in our work, faster in our travel, and proficient at leisure activities. These inventions came about as a result of using the materials found in nature and through an understanding of the laws of physics. This has allowed us to fly several times faster than the speed of sound, record any type of music, develop and explore the Internet, and use the power of the atom to provide enough electrical power to supply the needs of cities with huge populations. Using the laws of nature, particularly those of physical science, to create devices is called *technology* (Figure 1–1).

Technology probably began when humans picked up stones and sticks for use as crude implements to dig in the earth or to use as weapons. Later they learned that the sticks and stones could be altered and shaped to provide a sharper edge or a better handle. As they became proficient at using these primitive tools, humans began to devise ways of making the tools better (Figure 1–2). At first, they simply used trial and error as a means of improving their inventions, but

Figure 1–1

Technology has allowed humans to develop many wondrous machines to make life better.

© Rainer Plendl/www.Shutterstock.com

Figure 1–2
Ancient people used technology to make simple tools such as this arrowhead.

somewhere in time, they began to study why matter and energy behaved in a particular manner. This began the study of the laws of physics. Such laws as those governing fire, gravity, friction, electricity, heat, leverage, and inertia slowly began to be understood.

Physical laws were put to use by combining materials and energy in the right way to produce different devices and machines. The study of physics has come a long way from the time when humans began using stones to the time when electronic and atomic energy was harnessed and used (Figure 1–3).

There are still many facets about physics that we do not understand, but humans have been able to make tremendous use of the laws we understand and even many of those we

Figure 1–3
The study of physics has come a long way since humans made the first simple tools.

do not. Just as we have pondered, studied, and understood the laws of physics, we are now in the age of beginning to understand the laws of biology. It might be argued that the laws of physics have been used more by scientists to understand and create inventions than the laws of biology that have been used by scientists to create changes in living organisms. Much of the reason for this is that higher-ordered living organisms are tremendously more complicated than machines governed by the laws of physics.

Imagine the most complicated of computers. It is impossible for most people to comprehend how these devices operate. Plug a computer into a phone line or a cable, and you can contact millions of people all over the world—it seems miraculous. Yet even the most advanced computer and network applications are simplistic compared to the operation of the body of a mammal or the functions of all the systems of a corn plant (Figure 1–4).

The principles behind the most simple of bodily functions have eluded the understanding of research scientists for many years. Since ancient times, we have known that animals breed to produce young and seed is planted to obtain new crops. Humans understood that young animals resembled their parents and that in order to get better animals, the best animals in the herd were mated. Seed from the best plants were saved for planting because it was known that the parent plants pass on the superior characteristics to the new plants through the seed (Figure 1–5). However, until relatively

Figure 1–4

The most complicated computer in the world is simple compared to the functions of a corn plant.

© iStockphoto/Jason Titzer

Figure 1–5
For thousands of years, humans have known that new plants came from seed. Seed from the best plants were kept and planted.

recent times, people had no idea how this process worked. Although many theories were formulated, it was simply referred to as "the great miracle of life." In the middle 1800s, Gregor Mendel developed the first theory of gene transfer that was usable to scientists. Over the years, this theory was used to make advances in plants and animals used in agriculture. Yet it was not until the latter part of the twentieth century that scientists began to understand how the process of gene transfer functioned.

The same example could be used with most of the functions of body systems of plants and animals. Scientists are just now unlocking the secrets of nature—and we are still far from completely understanding how living organism operates. When new breakthroughs are announced, the scientists usually state that the process is tremendously more complicated and intricate than they ever expected.

Unlocking the secrets of living organisms has been extremely difficult for several reasons. The development of mechanical and electronic technology was a process of applying a group of principles and laws that were already understood. For example, cars were developed because inventors understood the principles behind compression, ignition, torque, and gear reduction (Figure 1–6). Metallurgy, chemistry, and physics all contributed to the effort. The point is that scientists began with principles they understood and turned these principles into technology used to create our modern inventions. Quite the opposite is true of scientists dealing with the

© iStockphoto/James Steidl

Figure 1–6
Cars were developed when inventors began to understand the principles behind compression, ignition, torque, and gear reduction.

functions of living organisms. They must begin with systems that already operate in a tremendously complicated manner. Many times, each bodily function of an organism must rely on other functions to operate properly. For example, nutrient metabolism directly impacts everything from growth and energy to reproduction. To understand how growth takes place, a scientist must also understand how nutrient assimilation and metabolism operates and how they affect the process.

Through biotechnology, scientists and engineers work to improve agricultural systems, health care, and environmental sustainability. In order to harness biological systems and living organisms for our purposes, scientists have learned to modify basic processes inside cells through **genetic engineering**.

Defining Biotechnology

There are many ways to define **biotechnology**. One of the simplest definitions is that it is the manipulation of living organisms or parts of organisms to make products useful to humans. This definition generally includes working with a whole plant or animal and all the systems that make up the organism. Another way to define the term is that it is using knowledge of cells to modify their activities in order to make living organisms more effective in serving people. Biotech at this level deals with the cells of plants or animals and may even include cells of organisms, such as viruses, that are neither plant nor animals.

Still another definition of biotechnology is that it deals with the manipulation of the genes of organisms to alter their behavior, characteristics, or value. This type of biotech is by far the one with the most potential and is by far the most controversial. Throughout this text, we will explore all the definitions of biotechnology and how humans have developed technology for use with living organisms.

One reason the development of biotechnology has been so difficult is that scientists use living organisms that grow and reproduce (Figure 1–7). Also, in the case of higher-ordered animals, the care and welfare of the animals have to be taken into consideration. Manipulation of cells and cell function has to be done carefully with many safeguards to govern the process. This adds complications not encountered with most processes in the physical sciences. The public is very hesitant to accept organisms that were produced through means they consider to be unnatural. People are reluctant to accept anything they do not understand, particularly if it relates to the food they consume.

A large part of what is described as biotechnology is that of cell and gene technology used to produce new characteristics in plants or animals. This technology is used in a wide variety of ways to improve the plants and animals we

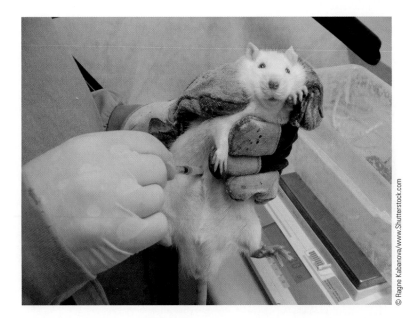

© Ragne Kabanova/www.Shutterstock.com

Figure 1–7
One reason the development of biotechnology has been so difficult is that scientists use living organisms.

Figure 1–8
Traditional breeding of plants and animals takes a tremendous amount of time.

produce for human use. The traditional breeding of plants and animals takes a tremendous amount of time because scientists have to wait for each new generation of plants or animals in order to observe the results (Figure 1–8). The desired result may not occur, and the scientists then have to wait for another generation. With genetically engineered organisms, scientists can place the particular gene that controls the desired characteristic in the precise place on the chromosome. In this way, the anticipated result can be achieved in the next generation of plant or animal.

Bioinformatics is the use of information technology to store, analyze, sort, label, and share the many DNA sequences generated by genome sequencing projects (Figure 1–9). In addition to DNA sequences, there are large collections of protein sequences, protein structures, metabolic pathway descriptions, research articles, patents, and many other types of biology-related data managed by bioinformatics scientists.

In the age of genomics, we have detailed information on the minimum number and type of genes required to support life. The knowledge of genomes and the tools of biotechnology have opened up new possibilities in genetic engineering. Some scientists are now working in the field of **synthetic biology**, where biological parts and entire living systems may be designed and synthesized in the laboratory.

Figure 1–9
Bioinformatics is the use of information technology to sort, analyze, label, and share the findings generated by biological research.

The Need for Biotechnology

Today there are over 6 billion people in the world. In the country of India alone, there are over 1 billion people. Projections are that by the year 2050, world population could approach 10 billion people. If these projections are close to correct, there is a lot of uncertainty about how the earth will support all these people. One thing is absolutely certain: all these people will have to be fed. Most of the increase in the world's population will come from the **developing countries** (Figure 1–10). Today, hundreds of millions of people in developing countries do not have an adequate supply of food, lacking the caloric intake needed to sustain body weight, fight infection and disease, or carry out light activities.

In the past, a large part of the population of developing countries lived in rural areas where they could produce at least a part of their food. Currently, the major portion of the projected population growth is expected to take place in the major cities of these countries, where there is no opportunity for people to grow a portion of their own food supply. This means that food will have to be produced and shipped into the cities at a rate greater than ever before.

During the past 50 years, tremendous advances have been made in the efficiency with which we grow food. Much of the increase in food productivity began in the 1970s, with what was termed the **Green Revolution**. During this time, developments in conventional crop breeding, new pesticides,

Figure 1–10
Projections are that by 2050, the world's population could approach 10 billion people. Most of the increase will come from developing countries.

Figure 1–11
The Green Revolution of the 1970s brought about large increases in yield of crops such as rice.

and management techniques started a dramatic increase in the amount of food produced each year. The greatest increase came about in rice and wheat, which are two of the world's leading food staples (Figure 1–11).

Since about the mid-1990s, statistics have shown that the yearly rate of increase for these crops is decreasing. Also there is evidence that productivity increases in other areas have slowed. For example, one of the staple food crops in many developing countries is a tropical root known as *cassava*. Since 1970, the amount of land devoted to the production of cassava has increased about 43 percent, while the amount of production increased by only 20 percent during that time. This is an indication that poorer quality land is being put into production. This has tremendous implications not only for feeding people but also for the impact on the environment.

It is apparent that the present rate of food production, coupled with the growing world population, means that many countries stand a real danger of running out of food in the near future. Crops (food, fiber, and feed), livestock (food, fiber, and labor), poultry, and aquaculture systems will need to be maximized for production, with care taken to minimize impact on the environment.

Biotechnology holds an enormous potential for feeding the world's population in an environmentally sustainable way. The technology exists to enable us to engineer plants and animals to produce more efficiently using fewer agricultural inputs, such as nutrients, water, or fertilizer (Figure 1–12).

Figure 1–12
Genetic engineering can produce crops that are higher yielding, cheaper to produce, and more nutritious.

Crops can be engineered to grow on poor soils in order to maximize land use without clearing more forests or wild habitats for farming. Better growth, higher yields, and shorter generations for both plants and animals can be achieved through genetic engineering. And it is possible to engineer plants and animals with quality traits, such as increased nutritional content or the ability to produce medically useful proteins, through biotechnology.

Current Uses of Biotechnology

Most scientists say that the future of agriculture is in biotechnology. Already, such biotechnology marvels as genetically modified organisms have made an impact on agriculture around the world. One of the economic advantages of biotechnology is that producers can use the technology without having to invest in additional machinery, equipment, land, or other capital outlay. The technology is "in the cell" (Figure 1–13). Outlined below

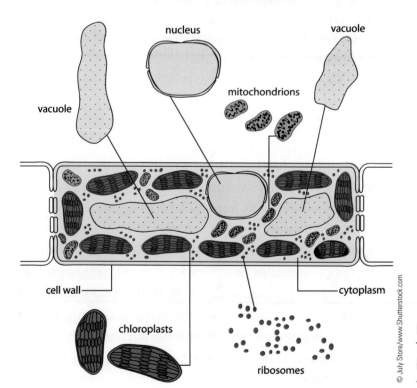

Figure 1–13
The "technology" of biotechnology is in the cell.

Figure 1–14

For many years, plants have been cloned using various techniques of asexual propagation.

Figure 1–15

Long ago, people discovered how to use enzymes and bacteria to make cheese.

are some of the current uses of biotechnology. More detail about these areas will be included in upcoming chapters.

Plant Biotechnology

One of the greatest impacts of biotechnology has come about in the area of plant agriculture. For many years plants have been cloned using various techniques of asexual propagation (Figure 1–14). More recently, plants that have been genetically modified by inserting or moving **DNA** have gained widespread use. These field crops are variously known as **transgenic**, **biotech**, **GM (genetically modified)**, or **GE (genetically engineered)** crops. Over 14 million producers in 25 countries planted and grew 330 million acres of biotech crops in 2009.

The most widely grown biotech crops are soybean, maize (corn), cotton, and canola. The biotech crops have been modified to resist insects and certain types of herbicides, increasing yields and making production easier. Other biotech crops have been modified to resist viral infections (pathogen resistance), produce useful proteins (plant-made products), remove heavy metals from soil and water (phytoremediation), tolerate harsh conditions (water use efficiency and nitrogen use efficiency), and to have increased nutritional content (quality traits). Table 1–1 illustrates the widespread use of biotech crops.

Animal Science

Biotechnology has been used for thousands of years in animal science. Early civilizations learned to make cheese and other food products from milk by using animal enzymes and bacteria (Figure 1–15). Later, the technology of artificial insemination was developed to help producers make rapid improvement in their herds. Later still, embryo transfer allowed genetic improvement from the maternal side of the herd. Scientists are now on the verge of making animal **cloning** a practical means of reproducing animals with superior genetics. Animal cloning is the process of making an exact genetic copy of an animal.

TABLE 1–1 Biotech Crop Production in 15 Countries (ISAAA, 2009)

Country	*Hectares planted (millions of hectares)[1]	Crops grown
United States	64	Soybean, maize, cotton, canola, squash, papaya alfalfa, sugar beet
Brazil	21.4	Soybean, maize, cotton
Argentina	21.3	Soybean, maize, cotton
India	8.4	Cotton
Canada	8.2	Canola, maize, soybean, sugar beet
China	3.7	Cotton, tomato, poplar, papaya, sweet pepper
Paraguay	2.2	Soybean
South Africa	2.1	Maize, soybean, cotton
Uruguay	0.8	Soybean, maize
Bolivia	0.8	Soybean
Philippines	0.5	Maize
Australia	0.2	Cotton, canola
Burkina Faso	0.1	Cotton
Spain	0.1	Maize
Mexico	0.1	Cotton, soybean

[1]A hectare is about 2 acres.

As with the plant industry, scientists are able to custom-design genetically modified animals that better serve the needs of humans. We now have farmed salmon that reach mature size in a fraction of the usual time (Figure 1–16). We also have goats and cows that produce therapeutic proteins in their milk, medical mice with humanlike immune systems, and pigs that produce phytase enzyme in their saliva, leading to low-phosphorus manure that is more environmentally friendly.

Medicine

One of the greatest impacts biotechnology has and will have on our lives is in the area of medicine, where many uses are already widespread. Therapeutic proteins, vaccines,

Figure 1–16
Because of biotechnology, we now can grow salmon that mature in a fraction of the usual time.

and other medicines may be produced in genetically engineered cell cultures through **biomanufacturing**, such as human insulin produced in bacteria. The cells of whole biotech plants and animals also may be genetically engineered to produce pharmaceuticals **(pharming)**.

Personalized medicine looks at the genome of individual patients to determine molecular causes of disease and to identify treatments that may be most effective (Figure 1–17). **Personalized genomics** is a new field that expands on traditional genetic testing. The difference is that we now have ability to sequence the entire genome of an individual rather than looking at a few genes at a time (Figure 1–18). **Gene therapy** has been used to treat

Figure 1–17
Personalized medicine looks at the genetic makeup of individual patients.

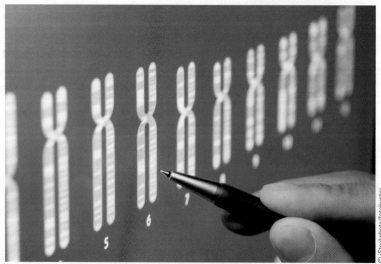

Figure 1–18
We now have the ability to sequence the entire genome of an individual.

and cure genetic disorders and illnesses that were once considered untreatable. Regenerative medicine now uses **stem cells** to repair and replace injured and diseased tissues. With these biotech tools, treatments for disease can be designed to maximize benefit and minimize side effects for patients.

The Environment

Biotechnology is used in monitoring and cleaning up the environment. Plants have been developed that can indicate whether or not an area is contaminated with pollutants. This allows for early detection of contaminants, and remedial measures can be taken before the problem becomes serious.

One of the most exciting ways that biotechnology can be used for the benefit of the environment is through the use of **bioremediation**, which is the use of living organisms to remedy an environmental problem (Figure 1–19). Certain microorganisms feed on toxins that are in polluted soil and water. These naturally occurring organisms actually digest the toxins and convert the pollutants to harmless substances such as carbon dioxide. Also, genetically engineered plants can be used to extract pollutants from the soil or water. Scientists are working to develop organisms that are more efficient at dissipating air, soil, and water toxins.

© iStockphoto/Baxternator

Figure 1–19
Bioremediation is the use of living organisms to remedy an environmental problem.

Summary

Humans have used biotechnology for thousands of years, since the dawn of early agriculture. However, it is only within the past 50 years that we have truly begun to understand how organisms live and reproduce. Unlocking the genetic code and understanding how traits are passed on from one generation to the next has opened almost unbelievable possibilities for improving our lives through biotechnology. Genetic engineering allows for the manipulation of organisms to benefit humans. Areas such as animal production, plant production, medicine, and environmental protection will continue to advance using the tools of biotechnology. In the next chapters of this text, you will discover some of the techniques, uses, and concerns of biotechnology.

CHAPTER REVIEW

Student Learning Activities

1. Search the Internet and make a list of all the food products that are currently produced using biotechnology. Compare your list to that of others in your class.

2. Choose a product from the list you made in exercise 1. Research how the product was made or processed and how the technology was developed. Report to the class.

3. Using your imagination, describe how a plant or animal could be modified by biotechnology in order to make life better for humans. Write a report that includes the description, how the technology might work, the benefits, and the dangers.

4. For a period of 2 weeks, look through the newspaper for articles dealing with biotechnology. Make a list of the points made in the article. Were the points negative or positive toward biotech?

5. Ask several of the teachers in your school to define biotechnology. How many different definitions did you get?

Fill in the Blanks

1. Using the laws of _____, particularly those of physical science, to create devices is called _____.

2. In order to harness _____ _____ and living organisms for our purposes, scientists have learned to _____ basic processes inside cells through _____.

3. Some scientists are now working in the field of _____ _____, where biological parts and entire _____ systems may be designed and _____ in the laboratory.

4. Through _____, scientists and engineers work to improve _____ _____, health care, and _____ _____.

5. _____ holds an enormous _____ for feeding the world's population in an _____ _____ way.

6. Biotech crops have been _____ to resist _____ and certain types of _____, increasing _____ and making production easier.

7. We now have farmed _____ that reach mature size in a _____ of the usual time, goats and cows that produce _____ _____ in their milk, medical mice with humanlike _____ systems and pigs that produce phytase _____ in their saliva, leading to low-phosphorus manure that is more environmentally friendly.

8. _____ _____ looks at the _____ of individual patients to determine _____ causes of disease and to identify _____ that may be most effective.

9. _____ medicine now uses _____ _____ to repair and _____ injured and _____ tissues.

10. _____ is the use of information _____ to store, analyze, sort, label, and share the many _____ sequences generated by _____ _____ projects.

True or False

1. Scientists have used both the laws of physical science and the laws of biological science equally in making modern advances in technology.

2. Genetic engineering is the use of information technology to store, analyze, sort, label, and share the many DNA sequences generated by genome-sequencing projects.

3. Living organisms are far more complex than machines governed by the laws of physical science.

4. One of the biggest challenges faced by the rapidly growing human population is how to feed people in developing countries.

5. Since 1970, the rate of production of cassava has increased more than the amount of land used to grow cassava.

6. The genetic engineering of crops and livestock can cause better growth, higher yields, and shorter generation times.

7. Biotechnology is more expensive than other types of agricultural technology because producers have to invest in additional machinery.

8. Biotechnology has been applied to animal science for the past few decades.

9. Animal cloning is the process of making an exact genetic copy of an animal.

10. Using bacteria to produce human insulin is an example of bioremediation.

Discussion

1. Provide at least two definitions of biotechnology.

2. Define genetic engineering.

3. Why have biological technologies been slower for scientists to develop than technologies based on physical sciences?

4. What was the Green Revolution?

5. What patterns do we expect to see in human population growth over the next several decades? Why is this growth pattern significant?

6. Describe at least two ways biotechnology was used by humans throughout history.

7. List at least four ways biotechnology is currently being used.

8. Describe how personalized genomics and gene therapy have improved health care.

9. Why is bioremediation such an exciting development for environmental technology?

10. In what ways might biotechnology be important for the survival of humans in the future?

CHAPTER 2

The Historical Development of Biotechnology

KEY TERMS

environment
plant breeding
enzyme
rennin
fermentation
yeast
heredity
law of segregation
law of independent
    assortment
cells
vaccine
antibiotic
artificial insemination
embryo transfer
genetic code
vertical gene transfer
DNA
double helix
genetic engineering
horizontal gene transfer
gene splicing

OBJECTIVES

When you have finished studying this chapter, you should be able to:

- List the important milestones in the development of biotechnology.
- Compare the development of agriculture with the development of civilization.
- Explain problems that were solved by the use of biotechnology.
- Analyze the importance of biotechnology in the development of food production and storage.

The Early Beginnings of Biotechnology

People generally think of biotechnology as being a new phenomenon; however, the use of technology to alter living organisms for human use began with early agriculture between 7,000 and 12,000 years ago. Before that time, humans lived a hunter-gatherer lifestyle, surviving by gathering fruits and seeds of plants they found in the **environment**. In addition, both large and small animals were hunted and killed for food. As useful plants and animals were depleted from an area, people would simply move their villages to places where food was abundant. Groups of hunter-gatherers transitioned to a settled lifestyle as humans found that they could contain and manipulate edible plants and animals, creating more reliable food supplies. This was the beginning of biotechnology (Figure 2–1).

As simple as it may sound, the selection of the very best plants and the saving of seeds from those plants was the

Figure 2–1
Biotechnology began when humans began to grow their own food.

© Peter Visscher/Dorling Kindersley RF/Getty Images

genesis of **plant breeding** and crop production. Humans also began to notice that some animals were more adaptable to domestication than others. As these animals were tamed, many uses were found not only for the meat of these animals but also for the hides, the hooves, the horns, and other body parts. For example, early humans discovered that sinews could be used as a very strong string for tying tools and other implements together. They found that by boiling hooves, a very strong glue could be made and used for binding animal skins together to make clothing and shelter. These simple acts were among the first processes of biotechnology.

Biotechnology and the Expansion of Civilization

Once humans could produce enough food to live in one area for an extended period of time, they had more time to develop labor-saving devices, and less people were required to produce food for survival. Individuals not involved in farming could then concentrate on developing new technologies, gathering and recording knowledge, and producing specialty goods. Early agriculture supported the development of villages and cities—civilization itself (Figure 2–2).

Figure 2–2
Early agriculture supported the development of villages.

As society developed, the people who produced food for the village found that growing plants and animals could be profitable. What first began as a barter system where animals and plant products were traded developed into a system where money exchanged hands. The hope of gaining something of value, as well as of producing food to eat, was a double incentive to find better and faster ways of producing food. Excess food could be traded to accumulate different items of value. Eventually, barter systems led to financial systems where money exchanged hands. Some anthropologists contend that perhaps this led to the development of money that could be more easily stored or carried than the larger items that had been previously traded.

Food Preservation

As trading systems expanded, people started to travel with their goods to distant places, bringing back goods from other peoples. Since traveling long distances necessitated storing food for the journey, people began to look for different ways to preserve food. There is a legend of a man in the Middle East who traveled across the desert with milk stored in the stomach of a calf. As people learned to use various products from the animals they slaughtered for food, the process of drying and curing the stomach of an animal was developed for use as a storage vessel for water and other liquids. The stomach pouch was tied to the saddle of a camel, and the liquid sloshed inside the pouch as the camel walked across the desert.

When milk was stored in the pouch the heat of the sun along with the sloshing of the milk caused the fat particles in the milk to coagulate. By the time the journey was ended, the milk had turned to cheese, and a new type of food preservation was created (Figure 2–3). Although the man did not understand the process or even know what made the milk coagulate, a form of biotechnology was used. We now know that an **enzyme** in the stomach of the calf called **rennin** started the process of coagulation that formed cheese. Since that time, the cheese industry has grown to include many hundreds of different types of cheese. Each year, thousands of tons of cheese are produced all over the world and provide a very nutritious part of people's diets.

Figure 2–3
Legend has it that cheese making was discovered when milk was stored in a pouch that was carried by a camel. The sloshing action turned the milk to cheese.

© iStockphoto/Photomorphic

© PzAxe. Image from BigStockPhoto.com.

Figure 2–4
Wine making was developed as a way of preserving grape juice.

As people began to settle in one place instead of roaming wide areas in search of food, they began to realize that in order to have food year-round, methods of storing food had to be developed. One of the first methods was the storage of fruit juice, such as grape juice. They noticed that as the juice began to spoil a change occurred in the juice (Figure 2–4). This process later became known as **fermentation**. They discovered that if the fermentation process could be halted

Figure 2–5
Someone discovered that adding yeasts to the mix would cause bread to rise into light, fluffy loaves.

at the correct time, the juice could be stored in a usable state. This was the beginning of the wine industry. As with the processing of cheese, new methods born from these discoveries led to newer and better products.

Bread Making

Another type of biotechnology that began early in civilization is that of making bread. The early humans noticed that many grass seeds are good to eat and that some grasses, such as wheat, produce an abundance of seed that can be ground into a powder called flour. Probably the very first bread that was baked from this powder had very little taste. Then someone discovered that by placing certain types of **yeast** into the mix the bread would begin to rise into light fluffy loaves that were quite tasty (Figure 2–5). Over a period of hundreds of years, humans have developed technologies and processes that created many types of bread, and each of these new developments involved biotechnology.

The Science of Genetics

For several thousands of years, humans gathered seed to plant so they could grow the crops they found in the wild. They merely selected those plants that produced the type of food they liked and planted seed from those that produced the largest quantity and the highest quality. They produced animals that were tamed from wild animals and, just as in plants, chose the best animals to reproduce. Slowly, they began to notice that different types of plants and animals could be bred to produce offspring that were superior to their parents. For example, they may have noticed that a slightly different type of wheat that grew in another area could be crossed with the wheat they were growing to produce a hybrid strain of wheat. A hybrid strain is a type of plant or animal that results from the crossing or mating of parents that are different. This was the beginning of plant breeding (Figure 2–6). However, it was not until the 1700s that people began to make a lot of progress in developing new varieties of plants, which was quite a feat considering they had no clue as to how traits were passed from one generation to the next.

In the mid-1800s, an Austrian monk named Gregor Mendel developed a theory of **heredity**, or how characteristics or traits are transmitted from parents to offspring. His theory was developed by watching and observing how plants grew in his garden. Of particular interest were garden peas, which he noticed had differences in the appearances of both the vines and fruit. For example, both the color and the texture of the seed varied from generation to generation. In addition, the stems of plants presented an interesting paradox. He noticed that plants with short stems could produce new plants with long stems and, conversely, that plants with long stems could produce plants with short stems. Through the observation of many generations of plants, Mendel developed an understanding of inheritance in pea plants (Figure 2–7). This understanding led to breeding research and programs which in turn led to the development of his **law of segregation**, which states that each parent provides one of the two forms of a gene for each particular trait. He also developed the **law of independent assortment**, which states that genes for certain characteristics are passed from parents to the next generation and are separate from the other factors or genes that transmit other traits. Mendel's laws were not widely known or understood until about 1900, when Hugh de Vries and Carl Correns independently rediscovered these laws of heredity.

Figure 2–6

Plant breeding began when people noticed that a slightly different type of wheat could be crossed with another type of wheat.

Figure 2–7
Genetic theory began when an Austrian monk studied how characteristics of garden peas were passed from generation to generation.

Figure 2–8
Scientists first studied cells during the seventeenth century when the microscope was invented. Contrast this one to a modern microscope.

The Discovery of Cells

Perhaps the earliest recorded milestone in the development of biotechnology came in the 1600s with the invention of the microscope. Up until this time, no one had even suspected the wonders of the microscopic world that could not be seen with the naked eye. In 1665, the scientist Robert Hooke removed a very thin slice of cork and examined it under the microscope. He observed tiny spaces that looked like the small rooms where prisoners in jails lived and called these structures **cells** (Figure 2–8).

As scientists began to examine more living tissue, they were amazed at the wondrous inner workings of living cells. Thousands of hours of study and research produced many theories about how cells operate. As the functions of living cells began to be understood, this knowledge opened the door for scientists to manipulate cells to create better products for humans.

Disease Prevention and Treatment

The health and well-being of both animals and humans were greatly improved in the latter part of the nineteenth century with the discovery that germs cause diseases. Up until that time, no one knew for sure why animals and people got sick. Many different theories were formulated, many of which seem silly in light of today's knowledge. Once germs were discovered, the means of treating and preventing diseases became a reality.

One of the biggest advancements occurred when a French scientist named Louis Pasteur developed a means of preventing animals from contracting diseases. He discovered that by injecting the blood from sheep that had contracted and survived the deadly disease of anthrax into healthy sheep, the disease could be prevented in the healthy animals. As with most great discoveries, other scientists took notice of the research and began to develop their own experimentations, which in turn led to the development of many new **vaccines** that helped create environments where animals could be raised disease free (Figure 2–9). In time, vaccinations were used on humans, bringing about a tremendous improvement in the lives of people. Dreaded diseases, such as smallpox, were prevented by an injection of vaccine. In fact, smallpox once devastated human populations, killing about

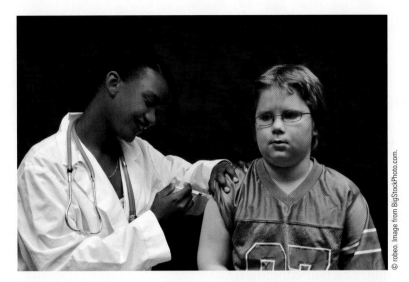

Figure 2–9
The development of vaccinations has brought about a tremendous improvement in the health of animals and humans.

Figure 2–10
Fleming noticed that no bacteria grew on mold cultures. This led to the discovery of penicillin.

80 percent of infected children and millions of people each year. Thanks to widespread vaccinations, this disease was certified as "eradicated" by the World Health Organization (WHO) in 1979.

In the late 1920s, the Scottish scientist Alexander Fleming was studying bacteria when one of his culture plates became contaminated with a mold called *Penicillium*. Just before he threw the culture plate out, he noticed that no bacteria were growing near the mold. He realized that the mold must be releasing a substance that inhibited the growth of bacteria (Figure 2–10). This began one of the greatest advances in the history of medicine. Extracts from the *Penicillium* mold were developed into the first **antibiotic** called penicillin. This led to several generations of so-called miracle drugs because of

Mystery Solved: Detecting the Source of a Boll Weevil Outbreak

In late summer of 2007, cotton growers in Texas's Southern Rolling Plains had reason to be alarmed. Hundreds of boll weevils were turning up in their fields. By the end of the year, more than 6,000 weevils were found in monitoring traps spread across several counties where the insect had been eradicated since 2000.

But a team of Agricultural Research Service scientists with a variety of skills was able to track down the likely origins of the reinfestation, giving cotton growers and entomologists some guidance on how to eradicate them and shedding light on just how far boll weevils can travel under certain conditions.

Boll weevils have been a threat to cotton growers since 1892, when they entered Texas from Mexico and proceeded to invade much of the south-central and southeastern United States. Large-scale eradication efforts, begun about 30 years ago, eliminated the boll weevil from much of the southern United States, but they remain a problem in pockets of eastern and southern Texas. Successful eradication efforts drastically reduce pest-management costs and insecticide use. Reinvasion is a constant threat, even in areas where they have been eliminated. Guarding against reinfestation is coordinated by the grower-supported Texas Boll Weevil Eradication Foundation, which funds monitoring programs and insecticide applications throughout the state. "Whenever eradication personnel capture boll weevils where they are not expected, they want to know everything possible about where the weevils may have come from. That information plays a major role in deciding the most appropriate response," says Thomas Sappington, an ARS entomologist in the Corn Insects and Crop Genetics Unit in Ames, Iowa. If the infestation is small scale, eradication personnel may need to spray only a few surrounding cotton

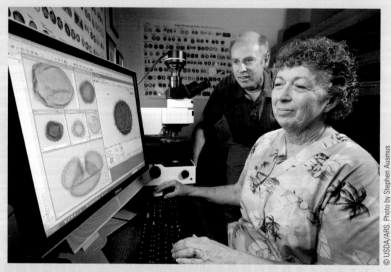

On a monitor attached to a compound light microscope, ARS palynologist Gretchen Jones and meteorologist John Westbrook examine images of pollen grains found on boll weevils.

fields. But if there is evidence of a widespread reinvasion, they may have to conduct more extensive spraying. In this case, the eradication foundation in the Southern Rolling Plains spent $1.4 million in increased insecticide applications alone, plus outlays for increased trapping, after they discovered the infestation.

Collecting the Evidence

Pinpointing the origins of the 2007 infestation took diverse skills. Sappington is an expert at using DNA to identify insect populations. Boll weevils travel on the wind, and John Westbrook, a meteorologist with the Southern Plains Agricultural Research Center (SPARC) in College Station, Texas, uses modeling techniques and weather data to analyze the effects of wind patterns on insect movement. Gretchen Jones, also based at SPARC, is a palynologist, or pollen expert, who can often identify an insect's itinerary by the type of pollen grains it picks up.

All flowering plants release pollen, and the pollen grains have distinctive shapes. Cotton pollen has a structure that makes it a particularly good forensic tool. It clings to the face, legs, and body of a boll weevil and will remain in the insect's gut for about 24 hours.

The researchers had one clue. Tropical Storm Erin swept through South Texas in August 2007, passing 112 miles to the south near Uvalde in an area known as the "Winter Garden District." Between Uvalde and the infested region is the Texas Hill Country, an arid region of scrubby vegetation with plants that produce types of pollen not found among the captured weevils.

Sappington focused on identifying patterns in the boll weevils' genetic makeup, comparing the DNA of 20 weevils captured in the reinfested Southern Rolling Plains zone with the DNA of dozens of weevils from 24 other sites in Texas and northern Mexico. Jones examined pollen grains found clinging to body parts and inside the guts of another 16 captured weevils, comparing them under a light microscope with pollen from plants blooming in Uvalde and in Cameron, an area east of the infested region where some officials suspected the weevils originated. Westbrook studied wind patterns and analyzed possible migration paths

A female boll weevil on a cotton boll.

with help from HYSPLIT, a computer model originally designed by the National Oceanic and Atmospheric Administration for federal studies that track the movements of smoke, particulate matter, and other airborne pollution.

The researchers concluded that the weevils likely came from the Winter Garden District. Westbrook found that Tropical Storm Erin skirted the southern and western sides of the reinfested region as it passed through, generating winds for 7 days that could have brought weevils up from the Winter Garden District. Jones found pollen on the invading weevils from nine plants common in Uvalde, including ragweed. She also found that the weevils lacked pollen from any of the numerous plants abundant in Cameron samples. Sappington reached the same conclusions by studying differences in DNA stretches known as "microsatellite loci," which can show collective patterns unique to each weevil population.

The results, published in the *Journal of the Royal Society Interface,* prompted growers to target the Winter Garden District for stepped-up eradication efforts. The work could also lead to better control measures. One way to prevent reinfestations is to harvest and destroy stalks early. The results of the study provide guidance on when approaching storms and hurricanes may warrant an early harvest. The multipronged approach also could be used as a model for resolving future questions about whether unexpected population spikes in fruit flies, aphids, or other pests are being caused by populations of insects from other areas.

—*By Dennis O'Brien, Agricultural Research Service Information Staff.*

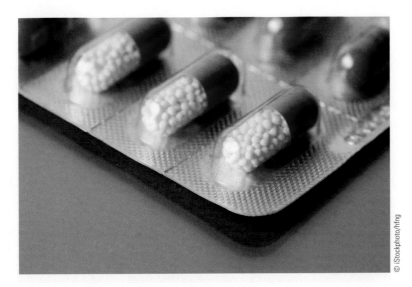

© iStockphoto/hfng

Figure 2–11
The discovery of penicillin led to the development of miracle drugs that have had a dramatic effect on bacterial diseases.

the miraculous effect they have on bacterial diseases. Antibiotics have saved millions of lives and have cured and eradicated many animal diseases (Figure 2–11).

Reproduction

Animal agriculture received a great boost when the technology of **artificial insemination** was perfected. According to legend, this technology was first used in the Middle Ages by Arabs who collected semen from stallions belonging to their enemies and bred their own mares to produce superior foals. The technology began to be used on a large scale in the United States in the 1930s (Figure 2–12). At that time, only fresh semen was used, which could be kept alive only about 2 or 3 days, and this limited the use of the technology. Then in the 1950s, adding a protective agent that allowed the semen to be frozen at temperatures reaching −320°F perfected the technique. This allowed the successful storage and shipping of semen all across the country and all around the world. For the first time, semen could be stored for long periods of time, causing very little harm to the sperm. In fact, bull semen has been successfully stored for as long as 30 years. This biotechnology greatly increased the availability of superior sires to small breeders (Figure 2–13).

Successful **embryo transfer** from one female to another became widespread in the 1970s. It allowed the use of superior females to produce multiple offspring, which, when

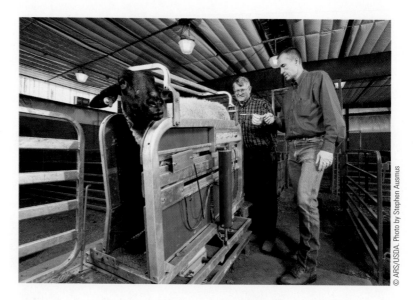

Figure 2–12
Widespread use of artificial insemination began in the 1930s. It has had a dramatic effect on animal production.

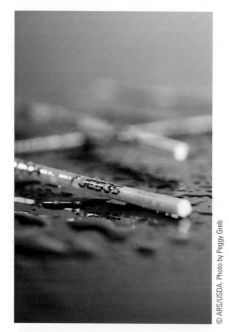

Figure 2–13
The addition of a protectant allowed semen to be frozen and stored in straws like this.

combined with artificial insemination, allowed very rapid progress in the production of superior animals.

Scientists have known for a long time that cells contain genetic materials capable of producing a new organism. Although scientists knew that it was theoretically possible to produce a complete new organism from a single cell, it was not until 1950 that this was accomplished. During that year, scientists were first able to grow a plant using the **genetic code** from a single cell using a process known as *in vitro* (meaning "in glass") culture. A new plant was started in a Petri dish rather than from a seed (Figure 2–14). From this technology, a process known as tissue culture was developed. Tissue culture technologies and other forms of plant propagation will be discussed in detail Chapter 8.

Gene Transfer

The transfer of genetic information from parents to offspring is called **vertical gene transfer**. Mechanisms of vertical gene transfer were almost a complete mystery until the 1950s. Although scientists knew that the nucleus of a cell contained the material for genetic transfer, the process was not understood. Then, in 1953, two scientists by the names of James Watson and Francis Crick published a model of the structure of deoxyribonucleic acid (**DNA**), the genetic material found within the nucleus of cells. They discovered that

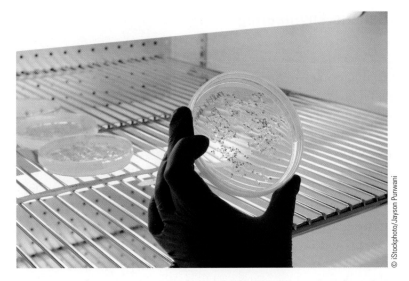

© iStockphoto/Jayson Punwani

Figure 2–14
Tissue culture allows the development of plants from a single cell.

© Jason Reed/Photodisc/Getty Images

Figure 2–15
The idea of genetic engineering began when the helix-shaped spirals of DNA were discovered.

DNA is composed of alternating units of phosphoric acid and deoxyribose and is formed in the shape of a **double helix**. The number, order, and type of nucleotides that form the DNA of this double helix determine the genetic code transmitted by the DNA. Nucleotides are the molecules that link together to form strands of DNA. These two scientists, analyzing a large body of evidence produced by other scientists, found that the entire key to the transmission of hereditary information from one generation to the next was contained within the double helix (Figure 2–15). Watson and Crick, along with collaborator Maurice Wilkins, received the Nobel Prize for their work in 1962.

Genetic Engineering

Although the discoveries by Watson and Crick were monumental in piecing together the puzzle of how genetics work, the real application for the knowledge and process was put to use in the 1980s. During this time, scientists discovered how to transfer bits of genetic information from one organism to another, allowing the expression of desirable traits in the recipient organism in a process now known as **genetic engineering.** It turns out that many microbes are also natural "genetic engineers," transferring genetic material between cells via processes of **horizontal gene transfer.** The original concept of gene transfer between parents and offspring has been adapted to include the transfer of genetic material from one organism to another organism through natural processes and biotechnology. For the first time, humans were actually able to remove a gene from one organism and successfully transplant it into another. This process, known as **gene splicing**, opened opportunities in an unbelievably wide array of applications. To many people, the term *genetic engineering* has become synonymous with the term *biotechnology*. Indeed, genetic engineering is a very large part of biotechnology; however, many new technologies and breakthroughs in the use of living organisms do not rely on genetic engineering. Throughout the remaining chapters of this text, biotechnology discoveries, developments, uses, and concerns will be thoroughly discussed.

Summary

Biotechnology is almost as old as civilization itself. Humans have always made use of the organisms they found in nature for food, shelter, and clothing. Throughout the past thousands of years, new discoveries and developments have helped us to better understand how living organisms live and reproduce. The more we understand about these processes, the more we are able to use biology and technology to improve our lives. Our diets, health, and comfort have all been greatly improved through the developments of biotechnology. Even though we have accumulated a lot of knowledge about how organisms live, grow, and reproduce, we have only just begun to understand these processes. Someday the exciting new discoveries we see today will be considered old technology, and processes we cannot even imagine will excite a new generation of people.

CHAPTER REVIEW

Student Learning Activities ....

1. Determine what you consider to be the most important milestone in the development of biotechnology. Compile a list of reasons why you consider the development to be the most important and share your reasons with the class.

2. Determine what you consider to be the most important biotechnology discovery of the future. List why you feel this technology is needed and explain what problems it could solve. Share your thoughts with the class. Try to predict when the discovery or technology will be developed and keep a list of all the class members' predictions for future reference.

3. Interview your grandparents or other elderly persons in the community. Ask them to share their ideas on what they consider to be the most important advancement in biology in their lifetime. Report to the class.

Fill in the Blanks ....

1. The _____ of the very best _____ and the saving of _____ from those plants was the genesis of _____ _____ and crop production.

2. We now know that an _____ in the _____ of the calf called _____ started the process of _____ that formed cheese.

3. The early humans noticed that many _____ _____ are good to eat and that some grasses, such as _____, produce an abundance of seed that can be ground into a _____ called _____.

4. In the mid-_____s, an Austrian monk named _____ _____ developed a theory of _____, or how characteristics or _____ are transmitted from parents to _____.

5. Pasteur discovered that by _____ the _____ from sheep that had contracted and _____ the deadly disease of _____ into _____ sheep, the disease could be _____ in the healthy animals.

6. Dreaded _____, such as _____, were prevented by an _____ of _____.

7. They discovered that _____ is composed of _____ units of _____ _____ and _____ and is formed in the shape of a _____ _____.

8. Embryo transfer allowed the use of _____ females to produce _____ offspring, which, when combined with _____ _____, allowed very _____ progress in the production of superior animals

9. In the 1980s scientists discovered how to transfer bits of _____ information from one _____ to another, allowing the _____ of _____ traits in the recipient organism in a process now known as _____ _____.

10. Our _____, _____, and _____ have all been greatly improved through the developments of _____.

True or False

1. Biotechnology is a relatively new phenomenon.

2. Agriculture contributed to early bartering and money exchange systems.

3. One of the first methods for storing food was by producing fruit juice and then fermenting the fruit juice into wine.

4. A purebred strain is a type of plant or animal that results from the crossing or mating of parents that are different.

5. Gregor Mendel observed his pea plants and found that short-stemmed plants produce only more short-stemmed plants and that long-stemmed plants produce only more long-stemmed plants.

6. Alexander Fleming discovered the antibiotic penicillin accidentally when his experiments with bacteria became contaminated with mold.

7. Bull semen has been successfully stored for as long as 30 years.

8. Microbes can function as natural genetic engineers and can complete horizontal gene transfer.

9. The first plant was grown in vitro in the 1930s.

10. Genetic engineering is the only type of biotechnology that scientists are currently using and studying.

Discussion

1. Explain how agriculture supported the development of cities and villages during early human civilization.

2. Describe Mendel's law of independent assortment and Mendel's law of segregation.

3. List at least three problems of human civilization that have been solved by biotechnology.

4. Who discovered the cell? What technological advance made this discovery possible?

5. Why are antibiotics considered "miracle drugs"?

6. What do you think it means to say that smallpox has been "eradicated"?

7. What are the benefits of artificial insemination and embryo transfer?

8. What is vertical gene transfer?

9. What did Watson and Crick discover? Why was this discovery important?

10. Describe the process of genetic engineering.

CHAPTER 3

The Principles of Scientific Research

KEY TERMS

research
basic research
applied research
Hatch Act
experiment stations
patent
intellectual property (IP)
scientific method
statement of the problem
hypothesis
theory
literature search
fact
experimental design
reliability
validity
treatment group
control group
data
t-test
significance
confidence level
correlated
peer review

OBJECTIVES

When you have finished studying this chapter, you should be able to:

- Define research.
- Discuss why research is so important to our way of life.
- Define patents and intellectual property (IP) rights.
- Explain the steps in the scientific method.
- Distinguish between a hypothesis and a theory.
- Explain how an experimental design is organized.
- Discuss how experiments are conducted.
- Explain how data are analyzed.
- Explain how research results are reported.

Scientific Research

Americans enjoy a higher standard of living than any other people who have ever lived. Never in history have any other people enjoyed the riches that we in the Unites States have at our disposal. Scientific endeavor has made our lives more comfortable and convenient than at any other time in human history. Most of us have an abundance of food and live in very comfortable houses; we have myriads of modern electronic conveniences, and we have many medical marvels that help keep us healthy. Most of the things we need in life are readily available, and this is particularly true for food, shelter, and clothing. The high standard of living that modern humans enjoy is to a great degree a direct result of scientific research, technological advances and human ingenuity (Figure 3–1). As our knowledge base has expanded over the years, both the quality and the quantity of our food supply have increased dramatically, whereas the relative cost of food has gone steadily down. Through properly applied scientific research, we have achieved levels of understanding about the physical and natural worlds that were only dreamed of by past generations.

In its broadest sense, **research** is the creation of new knowledge and usually involves gaining an understanding of natural phenomena. Generally, there are two types of research—basic and applied. **Basic research** deals with the investigation and understanding of how nature functions. In the biological sciences, this includes the study of how organisms

Figure 3–1
Research has resulted in the high standard of living we enjoy today.

live, function, grow, and reproduce (Figure 3–2). For example, basic research may investigate how a plant cell divides and reproduces. **Applied research** relies on the discoveries made in basic research and finds ways to make use of the knowledge. The basic knowledge of how a plant cell divides and reproduces can be put to use by scientists who discover how to create new plants from a single cell. This new knowledge can then be used by producers to propagate plants (Figure 3–3).

© mfoz. Image from BigStockPhoto.com.

Figure 3–2
Basic research deals with the investigation and understanding of how nature functions.

© Stevanovic Igor/www.Shutterstock.com

Figure 3–3
Applied research finds ways to use the knowledge found from basic research.

Figure 3–4
Experts in a research area meet at conferences to critically review each other's work and to share technological advances.

Most modern scientific breakthroughs come about as basic knowledge accumulates and different research groups add to the existing body of information on a topic through the publication of research papers. Usually, discoveries and innovations are brought about by teams of scientists who specialize in a particular area of study. The experts in a research area sometimes have different theories and meet at research conferences to discuss new ideas, critically review each other's work, and share technological advances (Figure 3–4). Scientists must communicate their findings to other scientists and to the world at large in order for their work to be useful for humanity.

During the past two centuries, individual scientists, as well as research teams, have made wonderful breakthroughs and innovations. Currently, the power of the Internet and the open science movement have increased science communication across all groups, making our shared knowledge more transparent and searchable than at any other time in history. Often, many years of painstaking work and data collection are required to verify facts before they may be considered part of the scientific standard or common knowledge base. The excitement of science is found when many small pieces of the puzzle come together to give us a clearer understanding of the natural world. Biotechnology uses our understanding of living things to develop useful products, processes, and services for humanity.

Current opportunities for individuals and research groups to synthesize scientific knowledge, correctly design experiments, and make new discoveries are better than these ever have been, but that has not always been the case. Centuries ago, research and development was conducted primarily by trial and error (Figure 3–5). More often than not, discoveries were made entirely by accident. This type of undirected research did not make the best use of resources and limited the acquisition of new knowledge. In 1862, the United States Congress established land grant institutions throughout the country to teach practical applications such as mechanics and agriculture. The biggest problem encountered with teaching agriculture was a very limited amount of knowledge of how to grow crops and produce animals. Over the thousands of years that agriculture has been practiced, many attempts have been made to explain the best practices to be used in growing food. Many superstitions and myths related to planting, tending, and harvesting crops and to the practice of animal husbandry are found in ancient societies. Agriculture arose independently in the Middle East, the Far East, and Mesoamerica between 7,000 and 12,000 years ago, before humans used writing to record knowledge.

Ancient people often did not understand the natural laws underlying their observations and experiences of agriculture. They created rationales and methods based on observations. Some of these reasoning were based on facts; however, many were myths and superstitions. For example, in northern Europe, lucky bonfires were lit by early farmers in spring and summer to ensure a good harvest or production of livestock. Until relatively recently, many farmers watched the phases of the moon to determine the best time to conduct management practices (Figure 3–6). One such myth was that corn planted during a full moon would have shorter stalks at maturity. Another common belief was that blemishes and bruises on apples picked during a waning moon would dry up rather than causing the fruit to soften.

These explanations seem silly to producers today; however, at the time, most people thought that these myths were based on fact. To address these types of superstitious agricultural practices and discover the natural laws governing plant and animal husbandry, the **Hatch Act** was passed in 1887. This congressional act established agricultural **experiment stations** operated by the land grant colleges. The idea was to

Figure 3–5

Long ago, scientific research was not conducted systematically, and many discoveries were made purely by accident.

Figure 3–6

Many years ago, farmers used astrology and the phases of the moon to determine when to conduct management practices.

Figure 3–7

Congress established experiment stations to conduct research within different areas of each state.

conduct scientific research in every state and even in different areas within each state (Figure 3–7). Since their establishment, the experiment stations have benefited all of society on an enormous scale.

Through the knowledge generated by scientists employed by our land grant universities, producers have been able to raise animals and grow crops many times more efficiently than ever before. For example, the development of hybrid corn varieties increased yields dramatically within just a few years. Selective breeding and a more thorough understanding of the nutritional needs of agricultural animals allow producers to supply the population with meat grown during a much shorter period and at a much lower cost. These and many other advances brought about by experiment station research have given us the most abundant, inexpensive, and safe food supply in history.

Today, many large companies conduct their own research to increase yields and provide consumers with high-quality products. Agricultural companies have developed many of our most useful varieties of seed, pesticides, pharmaceuticals, machinery, and processes (Figure 3–8). Agricultural technologies developed by companies are almost always subject to **patent** protection, which prevents competitors from selling related products while the company realizes a return on its research investment. Land grant universities also patent useful inventions and improvements.

Figure 3–8
Agricultural companies have developed many useful varieties of seed, pesticides, pharmaceuticals, and machinery. Most of their work is patented.

Figure 3–9
Universities and industry may share the cost of conducting research.

As land grant universities and agricultural companies develop collaborative projects, there has been increased focus on making sure that research results will benefit the public good. Companies may supply money for university scientists to conduct research with their products or to develop new products. Also, universities and industry may share the cost of hiring scientists, equipping labs, and conducting research (Figure 3–9). These collaborative efforts are often very efficient means of discovering new knowledge, though the legal frameworks around ownership of **intellectual property (IP)** in the form of patents resulting from these projects must be carefully managed. Intellectual property refers to ideas, concepts, and inventions that individual scholars or groups of scholars create.

Intellectual property rights sometimes inhibit innovation and the development of new technologies if the owners of those rights cannot find ways to share information and develop new products. The use of intellectual property rights in the field of agriculture has become increasingly complex and controversial as biotechnology has developed. In particular, some scientists are concerned that publicly funded research conducted at land grant universities is providing a disproportionate benefit to companies that patent publicly available information. University and public sector scientists may then be prevented from using the agricultural technologies for the benefit of society. However, the emergence of modern biotechnology has brought about a new set of problems over the patenting of living organisms that have been genetically altered, a topic that will be discussed in depth in a later chapter.

In recent times, the land grant universities have partnered with industry to conduct research. Companies may supply money for university scientists to conduct research with their products or to develop new products. In addition, universities and industry may share the cost of hiring scientists, equipping labs, and conducting research. These collaborative efforts are often a very efficient means of discovering new knowledge.

The Scientific Method

To ensure that the findings are of value, scientific research should be done on a very concise, systematic basis. A systematic approach known as the **scientific method** has been generally adopted as the standard procedure for conducting research. The advantage of this approach is that all experimentation and research can be conducted consistently. In addition, any bias held by the researchers can be kept to a minimum.

The scientific method has seven different, very deliberate steps (Figure 3–10). By following these steps, scientists can make sure the results of all their experiments and research are consistent. Each one of the steps has to be followed in a

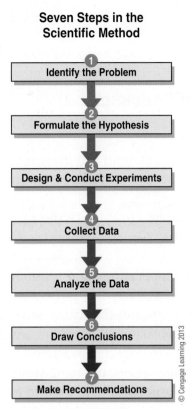

Seven Steps in the Scientific Method

Figure 3–10
The scientific method has seven steps.

very systematic manner because if one of the steps is improperly done, the whole research will be questionable. The steps that must be followed are stating the problem, formulating a hypothesis, designing the research methodology, conducting the experiment, collecting and analyzing the data, drawing conclusions and implications, and writing the research report.

Stating the Problem

The very first step in any scientific research is to clearly define the problem. Unless a clear-cut, well-defined **statement of the problem** is formulated, the entire research project will lack direction. In order to state the problem, the research scientist must have a good idea of the why the research topic is important and worthy of the time, effort, and expense necessary to conduct a thorough research project. In other words, the statement of the problem provides a basis and rationale for conducting research. The problem must be stated in clear terms that indicate a question that needs to be answered (Figure 3–11).

ARS National Research Programs in Genomics and Genetics for Food Security

The Agricultural Research Service is a leader in developing and using genomic data to improve the development of agriculturally important animals, crops, ornamentals, insects, and microorganisms.

ARS's genomics research program is concentrated in three of the agency's national programs: Food Animal Production (NP 101), Plant Genetic Resources, Genomics, and Genetic Improvement (NP 301), and Plant Biological and Molecular Processes (NP 302). But genomics research has very broad applications, and research projects often involve extensive collaborations with other ARS national programs, such as Food Safety (NP 108), Animal Health (NP 103), Plant Diseases (NP 303), and Bioenergy and Energy Alternatives (NP 307). These and other national programs are described at www.nps.ars.usda.gov.

Sequencing genomes enormously expands our understanding as well as the number of genes that can be deployed to address aspects of better world food security and to increase sustainable food production. But such genome programs are too big and too expensive for any one agency—or even one country—to take on.

ARS continues to play a major role in forming the international committees and coalitions that select which genomes should be tackled next and ensuring that research tasks are complementary, not duplicative.

Because of the huge potential that genomics offers for improving crops, ARS has set a goal of developing genomic libraries with genotypic and phenotypic information for all accessions in the National Plant Germplasm System. This is a massive but accomplishable job.

On the livestock and poultry side, ARS is leading a major effort to use genomics to improve the efficiency of animal production, especially in the area of feed utilization, to help reduce costs for producers and consumers, and to reduce the environmental impact of agriculture. Research projects will also be using genome sequence data to develop a better understanding of the host-pathogen relationship for the most dangerous animal pathogens and to enhance our understanding of the immune response to enable improved vaccines and postvaccine technologies.

ARS plant and animal genomics programs are also coordinating the development of new informatics tools for management, collection, storage, retrieval, and analysis of the large data sets being generated by genomics. This coordination includes promoting the integration of "-omics" data with large-scale phenotypic studies and the development of software to incorporate genome-level data into national and international genetic evaluation programs that support standards of interoperability, data validation, and quality assurance; and also promoting accessibility of the published data.

The goal is to maximize accessibility, utility, and use of genomics data; avoid duplication; and leverage developments from other research communities. ARS is also promoting the development and evaluation of technologies for rapid assessment of genomic diversity to guide the choice of candidates for whole-genome sequencing.

Food security is an international issue. The research to put genetics and genomics to work to enhance food security is also an international effort, one in which ARS plays a leading role.

States Where ARS Does Plant Genomics Research

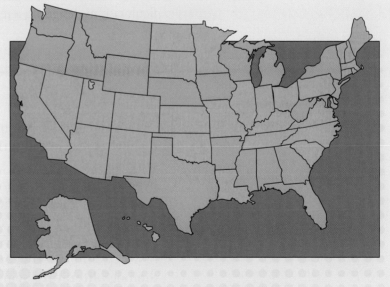

States Where ARS Does Animal Genomics Research

Published in the Agricultural Research *magazine.*

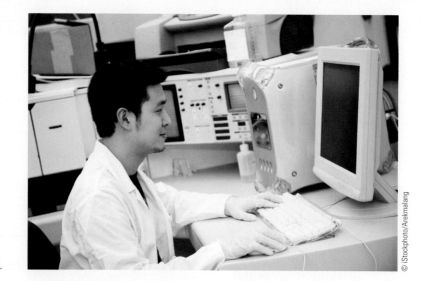

Figure 3–11
The first step in research is to write a clearly defined statement of the problem.

Scientists must also make sure that the problem they are investigating can be designed into a research study. For example, a study designed to determine if pigs have emotions could begin with a concisely worded statement of the problem in the form of a question: "Do pigs feel sadness and joy?" The statement is clear enough, but a scientist would have a very difficult task in designing a research study to answer the question.

Formulating a Hypothesis

Once the statement of the problem has been clearly defined, the researcher must construct a hypothesis. A **hypothesis** has often been defined as a **theory** or a guess, in the form of a statement, to a possible solution to a problem. However, a hypothesis is not a theory and is much more than just a guess because the researcher must have some basis for the hypothesis. A hypothesis can be developed based on observations or previously conducted research. This means that one of the first steps in properly conducting research is to carry out a thorough **literature search** for information on the research topic (Figure 3–12). By finding out what other scientists have discovered about the topic, the researcher is much better equipped to formulate a hypothesis. The hypothesis should state what the researcher believes (based on a study of relevant scientific literature) will happen as a result of the experimentation.

© iStockphoto/Andreas Reh

Figure 3–12
Researchers must conduct a thorough literature search for information on the research topic.

Some scientists prefer to write the hypothesis as a null hypothesis that states that there will be no differences in the groups when tested by the experiment. For example, a null hypothesis may be written as follows: There will be no difference in growth rate between catfish raised in water at 45°F and catfish raised in water at 65°F. If the scientist does not want to use the null hypothesis, the statement may be written as follows: Catfish raised in water at 65°F will have a greater growth rate than catfish raised in water at 45°F.

A hypothesis is not a theory or a fact. To a scientist, a theory explains existing observations based on the conclusions of several different research studies. Once scientists develop a theory, that theory may be constantly tested and may be rejected on further study. For example, for many centuries people theorized that the earth was flat. This was based on observations of many generations of people who had never traveled all the way around the world or had never known anyone who had traveled around the world. Once this theory was truly tested by the early explorers who circumnavigated the earth, the theory was rejected. On the other hand, a **fact** is a phenomenon that has been proven beyond a doubt. The law or theory that gravity holds everything on the earth is generally accepted, even though gravity is a force that has never been fully explained. However, it is a well-known fact that if a person jumps off the top of a building, he or she will fall to the ground. This phenomenon is a fact based on the observation that people or objects will always fall toward the earth without anything to hold them up.

Figure 3–13
Scientists must design a way to test the hypothesis. This is called the experimental design.

Designing the Research Methodology

Once a scientist has developed a hypothesis, he or she must then design a way to test the hypothesis. In the biological sciences, this is most often known as an **experimental design** (Figure 3–13). To a large degree, how well the experiment is designed and conducted determines the credibility of the study. Over the years, poorly designed studies have been conducted that have led to misunderstandings and misconceptions.

If you read news reports of recently concluded research, you may discover that there are often conflicting findings about the same topic. One report may claim that research indicates that a particular food additive may be harmful to a person's health, and another study may report the opposite. With all the conflicting information, how do you determine what findings are correct? The answer is in understanding how research is conducted and how to interpret the results of a research study. In order to use research results, the study has to be both reliable and valid.

Conducting the Experiment

Reliability and Validity

If other scientists are able to replicate a research study that has already been done and they come up with the same results, the original study is said to have **reliability**. In other words, how consistently will the experiment achieve the same results? If the experiment is repeated and different results occur, the research may not be reliable. On the other hand, if the experiment is repeated many times and the same results are achieved, the scientists can be reasonably sure that the research is reliable. Reliability has to be determined to ensure that the results of the study were not accidental or incidental.

Validity is a term used to describe whether or not a research study achieved what the researcher intended it to. A research study may be repeated several times with the same results and may be very reliable, but if the results do not measure what is really intended, the research is of little value.

In experimental research, validity is usually directly related to the researcher's ability to control all the conditions surrounding the experiment (Figure 3–14). Researchers generally use at least two groups of plants or animals they wish to study. One is called the **treatment group**, and the other is called the **control group**. In a well-designed study, both groups of plants or animals will closely match in characteristics. For example, if catfish are being studied, the fish in both groups should be the same strain and size, come from the same background, and be as much alike as the researchers can manage. One of the groups, the treatment group, will be given a treatment that the researcher intends to study. The control group will not be given the treatment to be studied but will be given the same feed, raised in identical tanks, kept

© iStockphoto/Andreas Reh

Figure 3–14
Validity is usually directly related to the researcher's ability to control all the conditions surrounding the experiment.

© Courtesy of ARS/USDA. Photo by Thomas Clarke

Figure 3–15
Both control and experiment groups must be given the same conditions and treatments except for the variable being studied. These are research plots.

under the same conditions, and otherwise managed in the same manner (Figure 3–15). If the two groups are different or treated differently in any way besides the treatment given to the treatment group, the research study may not be valid. This arrangement using a control and a treatment group is called an experimental design.

Collecting the Data

As the experimental design is formulated, the scientist must keep in mind what type of data is to be collected. **Data** are the bits of information obtained and recorded that will form the basis for accepting or rejecting the hypothesis. Data must be observable and most times must be quantifiable or measurable (Figure 3–16). In the example of the catfish experiment where we want to determine the rate of growth between the groups of fish raised in different water temperatures, the data would be relatively easy to gather and record. The scientist would periodically weigh the fish and record the weights or measure the fish and record the lengths. However, if a scientist wanted to determine which group of fish had the best taste when dressed and cooked, the data become a little more difficult to measure and record. In this case, the scientist must be innovative in ways to determine, measure, and quantify the differences in taste.

The collection of data must take place in a very consistent manner. Collection schedules have to be set and adhered to without fail. For example, if fish are to be weighed to determine differences in weight gain, they should be weighed at a certain time of day and at precise, regular intervals, such

© Goodluz/www.Shutterstock.com

Figure 3–16
Data must be observable and most times measurable.

© Razvan. Image from BigStockPhoto.com.

Figure 3–17
As soon as data are collected, they should be recorded in a journal.

as weekly or daily. When data are gathered, both the experiment and the control group data must be gathered in exactly the same way. The same scale, ruler, or other measuring device must be used. As soon as the data are collected, they should be recorded in a journal (Figure 3–17). Any condition that could possibly affect the data or the outcome of the experiment should also be recorded in the journal. For example, if the scientist finds four dead fishes in one of the tanks or if one of the tanks' aerators stops operating, the occurrence should be noted in the journal. Also, such conditions as nitrite, nitrate, and oxygen levels in each tank should be monitored and recorded in the journal.

Analyzing the Data

When all the data have been collected and the experiment is concluded, the data have to be analyzed. Raw data, as recorded, are generally of little use until properly analyzed.

Scientists use various statistical tools to test the data so that they can either accept or reject the hypothesis. For example, in the experiment we have discussed, the differences in growth rates of catfish raised in differing water temperatures were examined. Suppose that the scientist found that fish in the warmer water gained weight more rapidly than those raised in colder water. The determination must be made as to whether or not the same would be true of all catfish raised under similar conditions. Remember that the fish in the experimental group and the control group represent a sample of catfish. There is a possibility that the differences came about as a result of pure chance and had nothing to do with the water temperature. Therefore, the larger the number of fish in the samples, the less probability there is that the differences between samples happened by chance. In addition, using several treatment and control tanks can help to increase the odds that any differences were not by chance.

One tool used by scientists to analyze differences in samples is called a **t-test**. This test involves complicated mathematical calculations that are generally done by a computer. The data are fed into the computer, and the output comes in the form of a number called a *t*-value (Figure 3–18). This value is checked against a table that tells the researcher whether or not the difference is significant. **Significance** refers to the probability that the difference happened by sampling error or chance. The researcher usually decides before conducting the research what level of significance he or she is willing to accept. For example, is the researcher willing to accept or reject the hypothesis based on being 95 percent sure that the differences did not happen by chance? If so, the difference is said to be significant at the .05 level. This

Figure 3–18

Data are entered into a computer where a program analyzes the data and gives a statistical value, such as a *t*-value.

confidence level is quite common with research, such as that in our example of the catfish experiment. However, if scientists are testing a new drug for human use, being 95 percent sure that the drug is safe is not an acceptable risk—the data need to be significant at a much higher acceptance rate. If the acceptance rate is .001, this means that the researcher can be 99.9 percent confident that the differences were due to the treatment and not to chance. If the drug is tested many times and the same results are achieved at this confidence level, acceptance of the research is more likely.

Sometimes a scientist may not want to test the data for differences but may want to know if the data are **correlated**. Correlated data are related, and as one set of data increases, the other set either increases or decreases. If the researcher wants to know how water temperature affects the growth of catfish, he or she may do a series of experiments where fish are raised in several tanks with differing water temperatures. The scientist may find that as the water temperature rises, the growth rate also rises and that as the temperature is lowered, the growth rate is also lowered. The researcher can then say that water temperature is correlated to growth rate. Correlations are also tested to determine the likelihood that the correlation could have occurred by chance.

Drawing Conclusions and Implications

When the data are analyzed, the scientist must then draw conclusions based on the analysis (Figure 3–19). At this time, the hypothesis is either accepted or rejected. Care must be

© Radu Razvan/www.Shutterstock.com

Figure 3–19
When the data area analyzed, conclusions are drawn on the basis of the analysis.

taken not to go beyond reaching conclusions that are based on the data analysis. Almost always, the data merely show a correlation or differences and do not show a cause. For example, data can show that as more ice cream is consumed, the incidents of people drowning also increase. Although there is a strong correlation between the consumption of ice cream and drowning, the data do not show that eating ice cream increases the chance of drowning. While this may be implied, most people would realize that eating ice cream and swimming are activities that occur at the same time of year—the summer. Not all conclusions are as easily explained, and scientists must put a lot of thought into the findings of the data before they draw conclusions and make recommendations based on the research.

Writing the Research Report

When the research is completed, the scientist then writes a research report telling how the research was conducted and what the findings were (Figure 3–20). Included in the report are the rationale for doing the study, a statement of the problem, relevant information from the literature search, methods used in the study, data collection and analysis techniques, and conclusions and recommendations. Charts, graphs, and other illustrations are used to present the data in an understandable manner.

Figure 3–20
The last step in writing a report detailing how the research was conducted, the findings, and the conclusions.

The report is usually sent to a scientific journal for publication so that the results can be used by other scientists or anyone who has an interest in the findings. When the journal editors receive the report, they send it out to other scientists for review to make sure the research was properly conducted and reported and to ensure that only high-quality research is published. This process is called **peer review** and is one of the most essential parts of the research process. The research may also be presented at research conferences where other scientists listen to reports and scrutinize research.

Summary

Our lives have been greatly enhanced by the discoveries brought about as a result of scientific research. Myths and superstitions have been replaced by scientific facts, gathered in a large body of scientific knowledge. Researchers, working individually and in teams, communicate and critically review scientific knowledge as it accumulates. Over the years, scientists have developed a systematic approach to conducting research and accumulating scientific knowledge. This approach, known as the scientific method, is widely used to make certain that research is performed in a systematic manner that ensures the quality of the research. By following each step in the method, scientists can create knowledge that is valuable and dependable. By learning these steps, you can determine if the results of a study are sound. This knowledge can result in your making good decisions based on research.

CHAPTER REVIEW

Student Learning Activities

1. Go to the library or search the Internet for reports of research studies. These can usually be found in scientific journals. Locate a research study that interests you and report to the class. Explain how the study was conducted and what the researchers concluded.

2. Find a research study you consider to be flawed. Report to the class on why you think there are problems with the research and how you would correct these problems if you replicated the study.

3. Go to http://www.ffa.org and locate information about the Agriscience Fair. Design a research project for the fair. Pay close attention to the guidelines for this Career Development Event and design your project accordingly. Try to get the results published.

4. Choose an area of controversy in biotechnology, such as genetically modified corn used as human food. Determine what research has been done that supports each side of the controversy.

Fill in the Blanks

1. The high _____ of living that modern humans enjoy is to a great degree a direct result of _____ _____, _____ advances, and human _____.

2. Congress passed the _____ _____ in 1887, which established _____ _____ stations operated by _____ _____ colleges.

3. The emergence of modern _____ has brought about a new set of problems over the _____ of _____ _____ that have been _____ _____.

4. To ensure that the findings are of _____, scientific _____ should be done on a very _____, _____ basis.

5. A _____ explains existing _____ based on the _____ of several different research studies, whereas a _____ is a phenomenon that has been _____ beyond a _____.

6. In _____ research, _____ is usually directly related to the researcher's ability to _____ all the _____ surrounding the experiment.

7. _____ must be _____ and most times must be _____ or _____.

8. When the researcher determines if the _____ can be _____ or rejected, he or she must be careful not to go beyond reaching _____ based on the _____ _____.

9. When the journal editors receive a research _____, they send it out to other scientists for _____ to make sure the research was properly _____ and reported and to ensure that only high-quality research is _____. This process is called _____ _____.

10. The scientific _____ is widely used to make certain that research is performed in a _____ manner that ensures the _____ of the research.

True or False

1. Modern scientists usually make discoveries and progress working entirely by themselves.

2. Companies and land grant universities often work together to make the most out of scientific research and developments.

3. Intellectual property laws in agriculture and biotechnology are universally agreed on.

4. The main advantage of using the scientific method is that scientists can be sure the results of their research and experimentation are consistent.

5. A null hypothesis states that there will be no difference between the treatment group and the control group.

6. How well an experiment is designed and conducted often determines the credibility of a study.

7. In a valid experiment, the treatment group and the control group may have many differences.

8. If two sets of data are correlated, then we can prove that one set of data is causing a change in the other.

9. A significance level of .05 (95%) is always considered adequate.

10. Scientific research has greatly improved our standard of living.

Discussion

1. Explain the difference between basic research and applied research.

2. What does patent protection mean?

3. List the seven steps of the scientific method.

4. Why should scientists spend time formulating a well-defined statement of the problem?

5. Why is it inaccurate to say that a hypothesis is simply a guess?

6. What is reliability? What is validity?

7. How is it possible for two similar experiments to have drastically different results?

8. Explain the importance of using a journal while conducting scientific research.

9. Explain the effect of larger sample size on the significance of differences in results of an experiment.

10. Why is peer reviewing such an important part of the research process?

CHAPTER 4

Cells: The Foundation of Life

KEY TERMS

electron microscope
virions
prokaryotic cells
nucleoid region
deoxyribonucleic acid (DNA)
eukaryotic cells
cell membrane
diffusion
osmosis
homeostasis
cell wall
organelles
mitochondria
vacuoles
microtubules
microfilaments
ribosomes
endomembrane system
endoplasmic reticulum (ER)
lumen
Golgi apparatus
lysosomes
vacuoles
plastids
chloroplasts
leucoplasts
chromoplasts
gametes
meiosis
spermatogenesis
oogenesis
mitosis
interphase
prophase
metaphase
anaphase
telophase
totipotent cell
morula
stem cells
embryonic stem cells (ESCs)
adult stem cells (ASCs)
induced pluripotent stem cells
   (iPSCs)

OBJECTIVES

When you have finished studying this chapter, you should be able to:

• Explain why cells are the foundation of life.
• List the different types of cells.
• Distinguish between prokaryotic and eukaryotic cells.
• Describe the different components of cells.
• Explain how cells reproduce by mitosis.
• Explain the process of meiosis.
• Define and describe stem cells.

The Importance of Cells

The most basic life processes, such as reproduction, growth, disease immunity, and nutrient utilization, take place at the cellular level. All agricultural production begins with cells because producers depend on the healthy growth and reproduction of plants and animals for their livelihood. Not only is all of our food supply based on cell growth and reproduction, but without the proper functioning of cells, all life on earth would cease because new life is generated from existing cells that transfer life materials through a genetic code (Figure 4–1).

Almost all of modern biotechnology is based largely on the manipulation of cell functions and behaviors. We can use microbial, plant, and animal cells to provide useful products, processes, and services. Plants and animals are composed of many different types of tissue, and all tissue is composed of groups of cells. Each type of tissue serves a particular function, and to allow the tissue to serve its function, cells are specialized. In fact, such techniques as gene splicing and genetic engineering became possible only after scientists began to understand how cells function.

Remember from Chapter 2 that Robert Hooke was the first person to observe and describe the existence of cells. His observations were possible because of the invention of the microscope, that allowed him to see the cells that were

Figure 4–1

All of our food supply is based on the growth and reproduction of cells.

much too small to be seen with the naked eye. The type of microscope he used was a light microscope that made use of magnifying lenses and natural light. The problem with magnifying lenses is that very tiny objects, such as the components of cells, cannot be seen. Stronger lenses can be used, but the problem lies in focusing on the infinitesimally small objects. The stronger the lenses, the more difficult it is to focus light and the eye on the tiny object. Light microscopes allow for magnification of only about 2,000 times the original size of the viewed object.

With the invention and use of the **electron microscope** in the 1930s, scientists were able to visualize cell components and viruses for the first time. Throughout the 1940s and 1950s, the understanding of the anatomy and functions of cells was greatly increased using electron microscopy (Figure 4–2). Electron microscopes use beams of electrons rather than beams of light to illuminate microscopic objects. Magnetic fields in today's electron microscopes provide a focusing mechanism up to 2 million times magnification. The understanding of cell structures and functions gained

Courtesy of ARS/USDA

Figure 4–2

The study of cells was made possible by the invention of the electron microscope. This scientist is using a scanning electron microscope.

during this time frame laid the groundwork for biotech advances based on manipulation of cell division and the use of genetic engineering.

The electron microscope allowed scientists for the first time to examine all the parts of cells and to gain a better understanding of how these parts, or organelles, function. For the past 150 years, scientists had understood that cells were the fundamental basis of life, but they had little understanding of how cells functioned. The development of electron microscopes in the 1940s and 1950s opened a new world to scientists who were now able to look inside a cell and examine its components. Once they began to understand cell functions, such mysteries as gene transfer and cell reproduction began to be solved.

Types of Cells

Cells come in a wide variety of types and sizes. The smallest of cells are less than a micrometer (one-millionth of a meter) in diameter, and about the largest is an ostrich egg. Cells may be round like a ball, square like a box, long and thin like a string, or shaped like a plate, whereas others, such as an amoeba, have shapes that change constantly (Figure 4–3).

Cells are broadly grouped into two types: eukaryotic and prokaryotic. Both types of cells are similar in that they both contain genetic material and are filled with cytoplasm, a gel-like substance that is about 80 percent water and contains the cell organelles, proteins, the cytoskeleton, and other needed molecules. Within multicellular eukaryotic organisms, each type of cell has a particular role to play, and the shape of the cell is related to that role.

Viruses are not considered cells because they lack the ability to reproduce on their own (Figure 4–4). They must infect cells in order to hijack cellular machinery to translate their genetic code, sometimes causing obvious disease and destruction of the host cell. However, viral particles, or **virions**, have some similarities to cells in that they contain genetic material and structural proteins. Viruses infect a wide variety of prokaryotic and eukaryotic cells. The first virus discovered in 1898, tobacco mosaic virus (TMV), infects up to 125 different types of plants, including agriculturally important tobacco, tomatoes, and peppers.

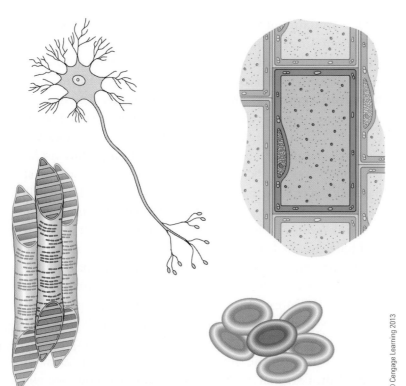

Figure 4–3
Cells come in a wide variety of sizes and shapes.

© Cengage Learning 2013

Figure 4–4
Viruses cannot reproduce on their own and are not classified as cells.

© Eraxion. Image from BigStockPhoto.com.

Prokaryotic Cells

Prokaryotic cells are the smallest of all cells, averaging 1–10 microns in size. Prokaryotes are the most numerous forms of life on Earth and make up two of the three

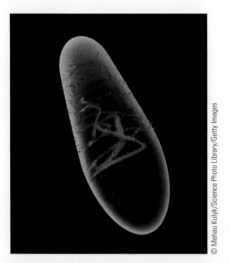

© Mehau Kulyk/Science Photo Library/Getty Images

Figure 4–5
The DNA in a prokaryotic cell is organized in a single molecule.

major domains of life, Archaea and Bacteria. Prokaryotic cells contain genetic material, but unlike eukaryotic cells, this material is not confined to a membrane-bound nucleus. Instead, prokaryotic genetic material is localized to the **nucleoid region** of the cell cytoplasm. The genetic material of prokaryotic cells (**deoxyribonucleic acid [DNA]**) is organized in a single molecule (Figure 4–5). Prokaryotes also lack membrane-bound organelles, though they do have ribosomes and cytoskeletal structures. Some prokaryotes also have self-replicating, extra chromosomal loops of DNA called plasmids, which may contain important genetic information.

Prokaryotes include one-cell organisms, such as bacteria and blue-green algae. We know a lot about prokaryotes that cause infectious disease in humans and animals, though disease-causing prokaryotes make up only a small fraction of all prokaryotes. One challenge in studying prokaryotic organisms is finding suitable conditions in the laboratory to grow the organisms. There are many more types of prokaryotic organisms in the environment than we can grow in a Petri dish. Fortunately, we are now able to sequence all of the DNA found in environmental samples, giving us a better idea of the types of organisms in the environment. This new field is called metagenomics and has shed light on the possible number of uncultured prokaryotic organisms waiting to be studied.

Prokaryotic cells easily exchange genetic information and were the first cells to be genetically engineered in the laboratory. The first recombinant pharmaceutical proteins, such as human insulin, were produced in *Escherichia coli* cell cultures (Figure 4–6). These cells are of tremendous value to biotechnology processes that have been around for thousands of years. For example, *Lactobacillus* bacteria play an essential role in the making of cheese and the fermentation of wine. Of course, the very earliest of cheese and winemakers were completely unaware of the use of these organisms, but they knew how to make the processes work. Later, as scientists began to understand the role microorganisms play in cheese and wine production, many advances were made.

In modern biotechnology, prokaryotic cells play essential roles in several areas. Prokaryotic cells are small, easy to grow in large quantity, and are relatively easy to genetically engineer. In the field of medicine, bacteria may be genetically engineered to produce human proteins, such as insulin,

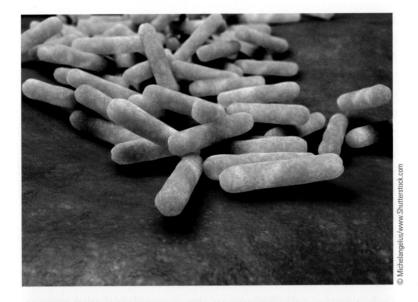

© Michelangelus/www.Shutterstock.com

Figure 4–6
Escherichia coli bacteria is used extensively in biotechnology. Insulin is produced by genetically engineered cell cultures.

© iStockphoto/Emesilva

Figure 4–7
Bacteria are used to manufacture a hormone used by cows to produce less fat and more milk.

or other useful drugs via biomanufacturing, which will be discussed in more detail in Chapter 12. Animal hormones useful in agricultural production, such as bovine somatotropin (also known as BST), may also be produced by biomanufacturing. BST is a naturally occurring hormone produced by the pituitary gland of cows that helps regulate the amount of nutrients a cow's body uses for fat or milk production. Supplementary amounts of BST cause the cow to produce less body fat and produce more milk (Figure 4–7).

Prokaryotes have been the "workhorses" of biotechnology since the 1970s. Much of what we know about molecular biology and genetic engineering was first discovered in research on prokaryotic cells. Prokaryotes are also the original source of endonuclease and ligase enzymes used to "cut and paste" specific pieces of DNA together, a critical set of tools for genetic engineering. Prokaryotes can easily exchange or share genetic information through horizontal gene transfer, where plasmids or fragments of DNA may move between cells. These natural processes inspired early genetic engineers. In Chapter 6, we will discuss how scientists have used *Agrobacterium tumifaciens*, an agriculturally important soil bacterium, to deliver genetic material into plant cells and make transgenic plants.

Eukaryotic Cells

Eukaryotic cells belong to the domain Eukarya, one of the three main branches of life. All plants and animals are made up of eukaryotic cells. Even though there are many differences between plant and animal cells, there are some similarities. As previously mentioned, all eukaryotic cells have a relatively large structure called a nucleus that serves as the control center for all activities of the cell, including reproduction. The nucleus is a membrane-bound compartment that houses the genetic material of the cell. The genetic material is called chromatin, composed of DNA and associated structural proteins. Chromatin is packaged tightly in structures called chromosomes (Figure 4–8).

The DNA code (packaged with proteins as chromatin and shaped into chromosomes) contains the information needed by the cell to build proteins and other nucleic acids needed for life. Along the DNA, information is organized into units called genes. Individual genes code for specific proteins and other useful nucleic acids, such as transfer ribonucleic acids (tRNAs) and ribosomal ribonucleic acids (tRNAs). As we will discuss in Chapter 5, genes are responsible for inherited traits.

The nucleus of plant and animal cells is surrounded by the cytoplasm. The cytoplasm contains membrane-enclosed structures called organelles that perform specialized functions within the cell. Both the cytoplasm and the nucleus are contained within the cell membrane. Some eukaryotic cells, like fungi and plants, also have a cell wall that surrounds the cell membrane, regulating cell shape and volume.

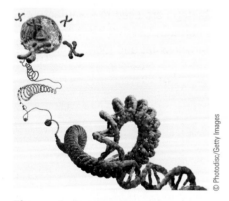

© Photodisc/Getty Images

Figure 4–8
Molecules of DNA are arranged in threadlike strands called chromosomes that are located in the cell's nucleus. Note the strands inside this nucleus.

Eukaryotic Cell Components

Cell Membranes

Every eukaryotic cell contains a **cell membrane**, also known as the plasma membrane, which serves three purposes. First, it encloses and protects the cell's contents from the external environment. Second, it regulates the movement of materials into and out of the cell, such as the taking in of nutrients and the expelling of waste. Third, the cell membrane allows interaction with other cells.

All material that passes into and out of the cell must go through the cell membrane. The membrane is selective, or semipermeable, which means that it allows only certain materials to pass through. The structure of the cell membrane is a phospholipid bilayer with embedded proteins. Not all substances are able to pass directly through the membrane. Only very small molecules, such as water, and uncharged (or hydrophobic) molecules, like some hormones, easily pass through the phospholipid bilayer. These molecules pass through the membrane in a process called diffusion (Figure 4–9).

Through **diffusion**, molecules in solution pass through the membrane from a region of a higher concentration of molecules to a region of lower concentration of molecules. For example, in an animal's cell, there are fewer molecules of oxygen inside the cell than there are outside the cell. In addition, there are usually more carbon dioxide molecules inside than outside. As the cell uses up oxygen molecules, more oxygen is allowed through the membrane because the molecules tend to try to equalize the number without and within the cell. Likewise, the carbon dioxide molecules move out of the cell to an area that is less concentrated with carbon dioxide molecules. Through the process of diffusion, the cell constantly takes in needed molecules, such as oxygen, and expels unneeded molecules, such as carbon dioxide.

Water is also passed through the semipermeable cell membrane in a process called **osmosis**. As in diffusion, the water moves from a region of high concentration to a region of low concentration, so the more material water has dissolved in it, the less concentrated the water is. If the cell has relatively little water inside, the solution tends to "draw" water from outside into the cell through the cell membrane (Figure 4–10).

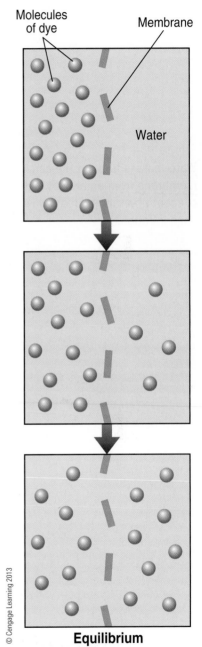

© Cengage Learning 2013

Equilibrium

Figure 4–9

Materials pass through a semipermeable membrane in a process called diffusion.

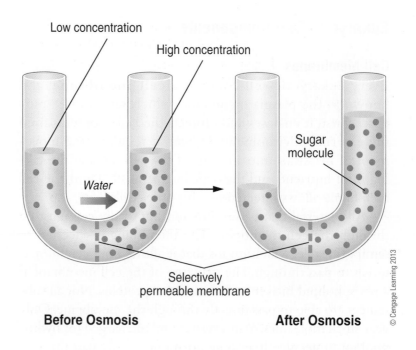

Low concentration

High concentration

Sugar molecule

Water

Selectively permeable membrane

Before Osmosis

After Osmosis

© Cengage Learning 2013

Figure 4–10

Water passes through the cell membrane from an area of high concentration to an area of low concentration.

Through the processes of diffusion and osmosis, the materials moving from one part of the cell to another and in and out of the cell are regulated. In all organisms, this is extremely important because these processes allow the cell to remain constant even though conditions in the environment may change. The ability of an organism to remain stable when conditions around it are changing is called **homeostasis**. For example, cells must retain the proper amount of water. The buildup of water in the cells creates an internal pressure called turgor that helps the cell retain its shape. When the cell is filled with the proper amount of water, the cells are filled out, and the membranes are taut. Cells in this condition are said to be turgid. In times of drought when plants do not have all the water they need, the cells lose their turgor and are limp and wilted. As soon as water is made available to the plants, the cells fill with water, the pressure builds, turgor is restored, and the plants are returned to their upright, healthy appearance.

Cell Walls

There are several differences between plant and animal cells. One of the most important differences is that plant cells have a structure called a **cell wall**. The cell wall gives the plant cell its shape and provides protection for the plant. It is composed primarily of a tough substance called cellulose.

Cellulose that provide rigidity for the walls of the cell and support for the entire structure of the plant. Cell walls were the structures Hook first saw when he examined cork under the microscope. Cellulose is very important in agriculture because such products as lumber, paper, and cotton are manufactured from the cellulose found in the cell walls of hard plant cells, such as those found in the branches and stems of plants (Figure 4–11).

Although cellulose is the key component of plant cell walls, they also contain hemicelluloses, pectin, and structural proteins. Hemicellulose provides a netlike structure that links cellulose microfibrils and pectin provides a gel-like cushion that fills the space between cellulose and hemicelluloses. Pectin is the component of plant cell walls that gives jams and jellies their thick consistency. Cell walls of the softer green plant parts, such as fruits and leaves, contain pectin. In tough woody plant parts, pectin is replaced by lignin. Lignin adds rigidity and strength to the plant tissue. The cell walls in woody plant tissues may also be composed of many more layers of cellulose, adding strength. Whether in green or woody portions of the plant, all cell walls have openings through which water, gases, and plant nutrients pass. Plant cells are held together by the contact of the individual cell walls.

Figure 4–11
Cotton is composed of cellulose which is made up of cell walls.

Organelles

Within the cytoplasm of all cells are structures that help the cell carry out its life functions. Prokaryotes have a couple of basic cell structures that are analogous to those found in eukaryotes, including ribosomes that produce proteins and a cytoskeleton for structural support and movement. In addition to these, eukaryotes also have membrane-bound organelles, such as mitochondria, endoplasmic reticulum, Golgi apparatus, lysosomes, and vacuoles (Figure 4–12). In much the same fashion as the organs of a body support an animal, **organelles** support the cell. One of the most important organelle is the peanut-shaped mitochondrion (plural, **mitochondria**), which functions to break down food nutrients and supply the cell with energy. Cells that use more energy contain more mitochondria than cells that are less active. For example, muscle cells contain more mitochondria than do bone cells because bone cells require far less energy than do muscle cells.

Vacuoles are organelles that serve as storage compartments for the cell. They consist of a membrane that encloses

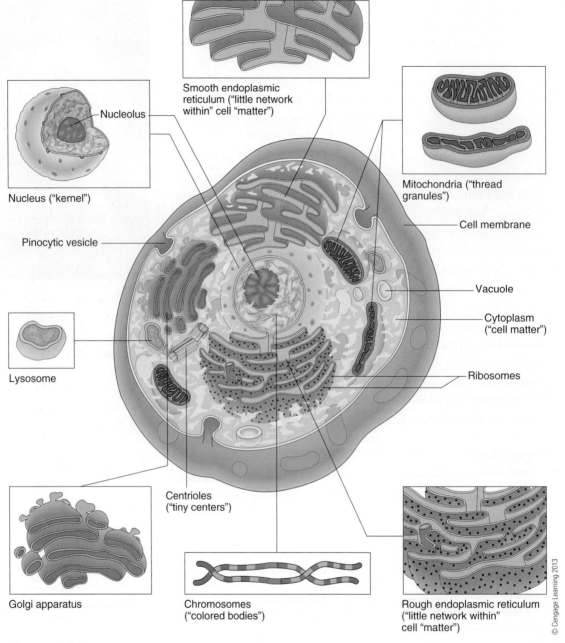

Nucleolus

Nucleus ("kernel")

Smooth endoplasmic reticulum ("little network within" cell "matter")

Mitochondria ("thread granules")

Cell membrane

Pinocytic vesicle

Vacuole

Cytoplasm ("cell matter")

Lysosome

Ribosomes

Centrioles ("tiny centers")

Golgi apparatus

Chromosomes ("colored bodies")

Rough endoplasmic reticulum ("little network within" cell "matter")

© Cengage Learning 2013

Figure 4–12
Animal cells contain many types of organelles.

water and other material. They store the nutrients and enzymes needed by the plants. Vacuoles also provide a storage space for the waste materials given off by the cell.

All eukaryotic cells have a cytoskeleton, composed of **microtubules** and microfilaments. Recently, prokaryotic

Figure 4–13
Microtubules are shaped like small thin tubes that give support to the cell and assist in the movement of chromosomes during cell division.

cells were found to have similar cytoskeletal structures. Microtubules are shaped like small, thin hollow tubes, composed of protein, that act as the "bones" and the transport highway of the cell. Motor proteins move vesicles and organelles along the network of microtubules within the cytoplasm. Microtubules also make up the mitotic spindle that moves chromosomes during cell division. Microfilaments work with microtubules to give support to the cell and are responsible for cell movements (Figure 4–13).

Microfilaments are fine fiberlike structures composed of protein. These organelles help the cell to move by waving back and forth.

Cells contain thousands of very tiny structures called **ribosomes** in their cytoplasm. These organelles are the sites where protein molecules are assembled in the cell. Proteins are needed by all cells for growth and other important functions. Enzymes that regulate the chemical process in the cell are composed of protein molecules.

The Endomembrane System

Transporting proteins and other needed molecules throughout the cell is a team effort that involves the nuclear membrane, the cell membrane, the endoplasmic reticulum, the Golgi apparatus, and vesicles that shuttle between these cell structures. These structures are all members of the eukaryotic **endomembrane system**. Proteins are passed through direct contact from one of these organelles to another or packaged in vesicles and moved along microtubules with the help of motor proteins.

BIOTECH in action

Marvelous Microbe Collections Accelerate Discoveries to Protect People, Plants—and More!

If you are a contact lens wearer, you probably remember headlines a few years ago about emergence of a worldwide medical problem—molds that live on contact lenses and cause debilitating eye infections.

What you may not have known: ARS experts at the National Center for Agricultural Utilization Research in Peoria, Illinois, did the detective work necessary to precisely identify these molds—which turned out to be *Fusarium* species.

These researchers derived the correct identification by working with a database of distinctive *Fusarium* genetic material that can be used to reliably differentiate among the many *Fusarium* species that cause disease.

In turn, this handy database owes part of its origin to the exemplary collection of hundreds of species of *Fusarium* housed at Peoria in the **ARS Culture Collection**.

Research leader and microbiologist Cletus Kurtzman and colleagues curate this comprehensive assemblage of living specimens of harmful and helpful bacteria, molds, actinomycetes (such as antibiotic-producing *Streptomyces*), and yeasts from around the planet.

Proximity to this gene bank—the world's largest publicly accessible collection of microbes—has, not unexpectedly, hastened discoveries by Peoria scientists. Their accomplishments include innovative new ways to detect, identify, classify (put in the correct family tree), and newly use these microorganisms to make foods safer, protect plants from pests, and create new industrial products.

Of course, other scientists also benefit. Some 4,000 strains of microbes are shipped each year from this flagship collection to researchers elsewhere.

Such sharing of specimens is all in a day's work at other specialized ARS microbe collections as well, including the **U.S. National Fungus Collections** in Beltsville, Maryland.

With more than 1 million dried specimens, these collections are the largest of their kind, according to director Amy R. Rossman. Scientists worldwide use them as a reference, and specimens are loaned for research projects.

Recent additions to the collections include seven species of fungi in the chestnut blight group that were

Microbiologist Cletus Kurtzman retrieves yeasts from the ARS Culture Collection.

discovered and described in the past few years. These descriptions will be used by forest pathologists to determine which species of fungi occur on hardwood trees, making it easier for these specialists to figure out how best to treat infected trees.

Coinvestigators Stephen Rehner, an ARS molecular biologist, and Joe Bischoff, a mycologist with USDA's Animal and Plant Health Inspection Service, have evaluated hundreds of isolates from the collection's holdings to make crucial molecular revisions of the taxonomy of two important fungal species. They are *Beauveria bassiana,* used to control termites, for instance, and *Metarhizium anisopliae,* which serves as a biological insecticide that controls termites, thrips, and grasshoppers.

ARS microbiologist Peter van Berkum, also at Beltsville, directs the **National *Rhizobium* Germplasm Resource Collection**. *Rhizobium* is the scientific term for bacteria that can form a symbiosis with alfalfa, soybean, and other legumes to provide these plants with a source of fertilizer—ammonia—for growth and reproduction. This is done by a process known as "biological nitrogen fixation"—the reduction of nitrogen gas from the atmosphere into ammonia. The collection provides a germplasm resource for industry to use in manufacturing inoculants and for researchers to use in investigating symbiosis and nitrogen fixation.

Current projects involving the collection include genetic mapping of *Rhizobium* found in association with alfalfa and other *Medicago* species from Egypt, Spain, and Tunisia. Also under way: Genetic mapping of *Bradyrhizobium* populations found in association with soybean. This investigation may determine—among other things—whether the *Bradyrhizobium* in United States soils are the same as those in the Far East, soybean's place of origin.

Both projects may lead to new ways to boost plants' productivity without using fertilizer. —By Marcia Wood and Alfredo Flores, Agricultural Research Service information staff. This research is part of Plant Diseases (303), Crop Protection and Quarantine (304), and Plant Genetic Resources, Genomics, and Genetic Improvement (301), three ARS national programs described at http://www.nps.ars .usda.gov.

Collections manager Erin McCray and mycologist David Farr examine a fusiform rust of pine, one of more than 1 million specimens in the U.S. National Fungus Collections.

The **endoplasmic reticulum (ER)** is a large webbing or network of membranes that is positioned throughout the cell and is a key part of the endomembrane system. The ER has an inner space or **lumen** where ions, such as calcium, are stored and proteins are folded and modified. Portions of the ER nearest the nucleus are dotted with ribosomes and is called the rough ER. Proteins destined for secretion or movement to the other cell locations are made in the rough ER, move within the lumen to the smooth ER, and are packaged in transport vesicles. The smooth ER also plays a role in lipid synthesis and detoxification. The **Golgi apparatus** in the cell is an organelle shaped like a group of flat sacs that are bundled together. Vesicles deliver proteins and lipids from the ER to the Golgi apparatus, where further modifications are made before secretion. Proteins and lipids within the Golgi are then repackaged in vesicles bound for the cell membrane, where they will be secreted outside the cell. One might say that the Golgi apparatus serves as the "post office" of the cell (Figure 4–14).

Lysosomes are organelles that serve as a cleanup crew within the cell, breaking down proteins, carbohydrates, and other molecules, as well as any foreign material, such as bacteria, that enter the cell. In addition, as other cell parts become worn out and nonfunctional, the lysosomes use their digestive enzymes to break these parts down. The products of the digestive actions are passed into the cytoplasm and out of the cell through the cell membrane.

Vacuoles are organelles that serve as storage compartments for eukaryotic cells. They consist of a membrane that encloses water and other material. Cells also store needed nutrients and enzymes in the vacuole as well as some waste materials. Plant cells typically have a large central vacuole that is responsible for maintaining turgor pressure within the plant cell.

Plant cells and algae have a class of organelles called **plastids** that are not present in animal cells. There are three types of plastids. **Chloroplasts** are plastids that use the energy of the sun to make carbohydrates. These organelles contain chlorophyll, which gives plants their green color. **Leucoplasts** are plastids that provide storage for the cell. They may contain starches, protein, or lipids (substances containing fatty acids). These organelles are abundant in the seeds of plants and contain nutrients that are used by the emerging plant or by animals that eat the seed. **Chromoplasts** are

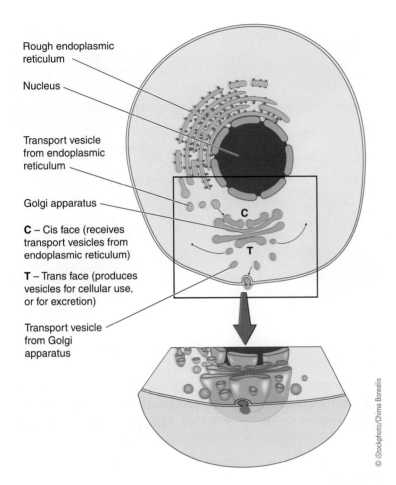

Rough endoplasmic
reticulum

Nucleus

Transport vesicle
from endoplasmic
reticulum

Golgi apparatus

C – Cis face (receives
transport vesicles from
endoplasmic reticulum)

T – Trans face (produces
vesicles for cellular use,
or for excretion)

Transport vesicle
from Golgi
apparatus

© iStockphoto/Chima Borealis

Figure 4–14
The Golgi apparatus serves as the
"post office" for the cell.

plastids that manufacture pigments that give fruits their color
and also give leaves their brilliant color in the fall when they
change color.

Cell Reproduction

The continuation of life depends on the reproduction of cells.
Even in the higher-ordered animals and most higher-ordered
plants, life begins with the uniting of cells, known as gametes,
from each of two parents (Figure 4–15). Once these cells have
united, the growth process begins, and cells multiply until an
entirely new plant or animal is formed. At both levels (reproduction and growth), the division of cells must take place.
When the gametes or sex cells are formed, a process known
as meiosis takes place. This process produces the sperm and
egg that unite to form an embryo. When the gametes have

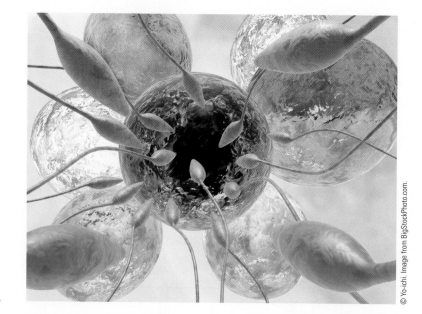

Figure 4–15
Life begins with the uniting of cells, known as gametes, from each of two parents.

merged through fertilization, the newly created cell begins to divide in a process known as mitosis. Through mitosis, growth and cell replacement take place. Biotechnologists must understand meiosis and mitosis to effectively use cells for the production of needed proteins or other purposes.

Meiosis

In order to form reproductive cells, or **gametes**, body cells must undergo meiosis. **Meiosis** is the process by which the normal number of chromosomes in each body cell is reduced by half. The normal number of chromosomes in any individual may be defined as 2n, where 1n came from the father and 1n came from the mother. The number of chromosomes varies between species. In humans, n = 23; in cattle, n = 30; and in pigs, n = 19. In other words you received 23 chromosomes (1n) from your father and 23 chromosomes (1n) from your mother, so your 2n number is 46 chromosomes. When a 1n sperm cell combines with a 1n egg, the combination of the two restores the 2n number of chromosomes, which is required for normal development of the organism. Remember that in sexual reproduction gametes from the male and female must join together to initiate the development of a new organism. Meiosis occurs in the testes in during the formation of sperm cells and in the ovaries during the formation of eggs. At the completion of meiosis, each human sperm cell will contain 23 chromosomes (1n), and each egg will contain 23 chromosomes (1n). When the egg is fertilized by the sperm, the 2n number of chromosomes (46) is restored.

The production of the male gamete or sperm takes place through a process called **spermatogenesis** that occurs in the testes of the male. Within these organs, cells develop into spermatozoa (sperm) through a four-step process (Figure 4–16). The first step involves a process called replication, in which the chromosomes make an exact copy of themselves. The replicated chromosomes are called chromatids. To complete the second step, the chromatids then come together and are matched up in pairs in a step called synapsis. In the third step, the cell divides, and the chromosomes are separated, with each

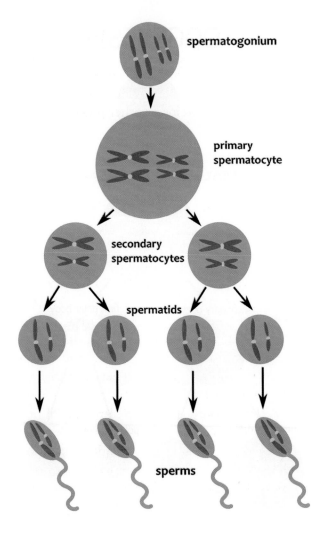

Figure 4–16
Sperm is produced through a process called spermatogenesis.

cell receiving one of each chromosome from each pair. However, remember that each chromosome replicated itself (chromatid) and is still attached together. In the fourth and final step, both the cells and the chromatids separate, and the chromatids become chromosomes. Remember that these cells (the new sperm cells) each contain only half the chromosomes that the original cell contained. The end result of this process is that four new sperm cells are produced from the original cell.

The female gametes (the eggs) are developed in the ovaries in much the same steps and processes as in the production of sperm. Gamete production in the female is known as **oogenesis** (Figure 4–17). There is, however, one important difference. In meiosis involving egg production, only one egg cell is produced, unlike the four new gametes formed in sperm

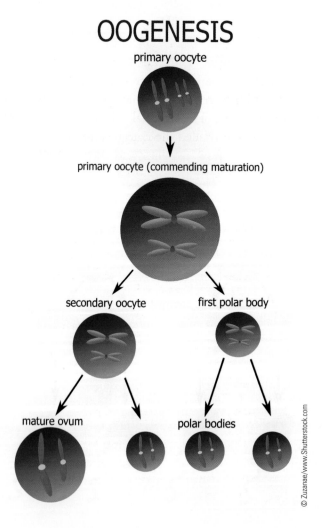

OOGENESIS

primary oocyte

primary oocyte (commending maturation)

secondary oocyte   first polar body

mature ovum   polar bodies

© Zuzanae/www.Shutterstock.com

Figure 4–17

Gamete production in the female is known as oogenesis.

production. In oogenesis, three of the newly divided cells become what are known as polar bodies, and only one cell becomes a viable egg. Since the egg is considerably larger than the sperm, it needs more nourishment. Most of the cytoplasm (cell material outside the nucleus) from the cell goes into the one cell that will become the egg. The function of the polar bodies is to provide sustenance for the egg until conception.

Mitosis

All of the growth that takes place in living organisms comes about as a result of cells increasing in size or numbers. The process by which these cells multiply is called **mitosis** (Figure 4–18). Cells have a very limited size to which they grow, so by far the greatest amount of growth in organisms comes about as a result of cells reproducing or multiplying. Also when injuries occur to either plants or animals, cells begin to reproduce in order to heal the wounds.

When a cell grows, it reaches a maximum size. When this size is reached, the cell divides into two cells. These cells in turn grow until they reach their maximum size, and each then divides into new cells. The original cell is called the parent cell, and the new cells are called the daughter cells. When a plant or animal matures, it stops growing, and the process of cell division is used to heal wounds and to replace worn-out cells.

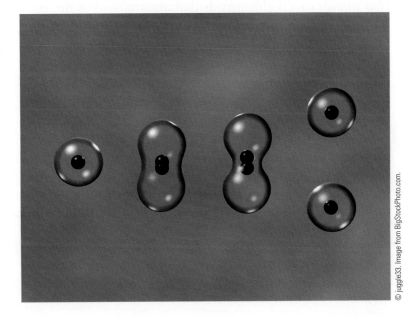

© juggle33. Image from BigStockPhoto.com.

Figure 4–18

The process by which cells divide and multiply is called mitosis.

Eukaryotic cells (cells that have nuclei) divide by mitosis. As mentioned earlier, all the genetic code for passing on traits of an organism is located in the nucleus of the cell. In the process of mitosis, all the genetic coding is duplicated and transferred to the new cells. Although the process of mitosis is continual, scientists have divided the events into different phases (Figure 4–19).

Interphase

The period when the cell is not actively dividing is called **interphase**. This phase is not really a part of mitosis but is a time when the cell is carrying on processes such as synthesizing materials and moving them in and out of the cell. It is during this time that the cell grows. As the cell reaches its maximum size, the DNA replicates and forms two complete sets of chromosomes. The threadlike molecules of DNA that make up the chromosomes are called chromatin and are spread throughout the nucleus. Animal cells have strands of genetic material outside the nucleus called centrioles. Most plant cells do not contain centrioles. At the end of the interphase, the cell is at the correct size, the chromosomes are duplicated, and the cell is ready to divide.

Prophase

The first phase of mitosis is called **prophase**. During this phase, the chromatin appears in the form of distinct, shortened rodlike structures. The chromosomes are formed of two strands called chromatids that are attached at the center by a structure known as a centromere. As this formation

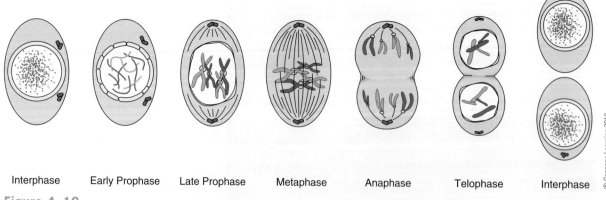

| Interphase | Early Prophase | Late Prophase | Metaphase | Anaphase | Telophase | Interphase |

© Cengage Learning 2013

Figure 4–19
Mitosis takes place through several separate phases. The process is continuous.

takes place, the nuclear membrane begins to dissolve, and the entire nucleus begins to disperse. In place of the nucleus, a new structure called the spindle is formed. The spindle is a structure shaped somewhat like a football and composed of microtubules that attach to the centromeres of chromosomes. In animal cells, centrioles play a key role in organizing the microtubules into the spindle structure.

Metaphase

During the next phase, called **metaphase**, the chromatids move toward the center of the spindle. The center of the spindle is referred to as the equator. When they reach the center, the centromeres of the chromatids connect themselves to the microtubules of the spindle.

Anaphase

Anaphase is the third stage of the process of mitosis. During this time, the pairs of chromatids separate into an equal number of chromosomes, and the centromeres duplicate. When separation occurs, the mitotic spindle moves the chromosomes toward opposite ends of the cell.

Telophase

The final phase of mitosis is called **telophase**. The chromosomes continue to migrate to the opposite sides of the cell (called poles). When they reach the poles, the remains of the spindle begin to disappear, and new membranes are formed around the chromosomes. This forms two new nuclei.

To complete the cell division, a process known as cytokinesis occurs. This divides the cytoplasm in the cell. Since mitosis is involved with the division of the nucleus of the cell, cytokinesis is a separate process from mitosis. In animal cells, a crease called a cleavage furrow begins to form in the center of the cell. This crease continues to deepen until the cell membrane divides along with the cytoplasm. One nucleus goes with each divided cell wall and cytoplasm, and the process of forming two new cells from the old cell is completed.

The cell walls of plant cells do not form a cleavage furrow like animal cells. Instead, a structure called a cell plate forms in the middle of the spindle and grows outward until the cell is divided into the two daughter cells.

At the completion of mitosis, the new daughter cells are genetically identical to each other and to the parent cell that

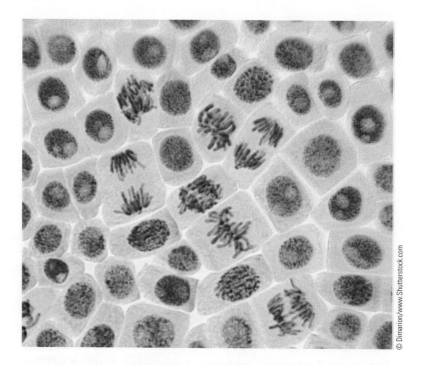

Figure 4–20

The process of mitosis is continual. See if you can identify different phases of mitosis taking place in these cells.

divided to form them. After formation, the daughter cells then go into interphase, and the whole process of mitosis starts over. Through this continuous process, an organism grows and maintains its structure through the replacement of worn-out and injured cells (Figure 4–20).

Animal Stem Cells

During sexual reproduction, once fertilization has occurred, a complete cell has formed with all the genetic material necessary for its development into a complete organism. This cell is said to be a **totipotent cell**, meaning that the cell is capable of developing into any type of cell that might be found in the body of a mature adult animal. Within a few hours, this cell divides into two totipotent cells, either of which could be implanted into a uterus and develop into a complete animal. Sometimes this happens naturally, and identical twins are formed.

Early embryonic cells divide and group together to form a ball-shaped mass called the **morula**, where the cells divide and clump into a mass in a process called cleavage. As the cells of the morula begin to increase, they form into a spherical shape called a blastula with an outer layer and an inner mass of cells. The outer layer of the blastula develops into the placenta, which attaches to the uterus and provides nutrients and other support for the fetus. The inner masses of cells form all the different types of tissues for the body.

As the blastula begins to grow and develop, the cells begin to change and take on different characteristics. The cells begin to form different layers that later develop into the organs of the body. Like cells group together to form tissue, and the tissue that develops bone is different from the tissue that develops blood, the tissue that develops muscles has its own unique characters, and so on. This process is called cell differentiation. The cells that begin the differentiation process are called **stem cells** (Figure 4–21). Changes in gene expression in the different types of stem cells trigger differentiation. Scientists are still working out the specific genetic and environmental signals that trigger gene expression changes leading to cell differentiation.

Embryonic stem cells may be useful in regenerating diseased and damaged tissues as part of biotech medical interventions, though the use of embryo-derived cells is highly controversial. Many researchers prefer to work with **adult stem cells**, which are found throughout the human body. These cells are not totipotent, but they are capable of developing into a number of different tissues that are useful in regenerative medicine. The other great benefit is that they

© Steve Gschmeissner/Science Photo Library/Getty Images

Figure 4–21
The cells that begin the differentiation process are called stem cells. These are embryonic stem cells.

would not trigger an immune response if derived from and returned to the same patient. **Induced pluripotent stem cells (iPSCs)** are stem cells that are derived from adult, differentiated cells through treatment with special proteins or viruses. These cells also hold great promise in their ability to regenerate healthy tissues, though work is in the early stages.

Ultimately, scientists working in the field of stem cells and tissue engineering would like to create new tissue to replace diseased or damaged human tissue, but first they must unlock the secrets of the cell differentiation process. Growing new tissues for a specific organ could potentially cure many diseases, such as Parkinson's disease, diabetes, heart disease, and other problems. If scientists are ever able to unlock the secret to the cell differentiation process, the possibilities will be enormous. The use of human stem cell research is extremely controversial, and until issues surrounding using these cells are settled, the research cannot proceed unhindered. This topic will be covered more in depth in a later chapter.

Summary

Cells are the basic building blocks of life because all the processes involving growth, reproduction, and maintaining the well-being of the organism happen at the cellular level. This is why a comprehension of how cells function is so crucial to the understanding of biotechnology. The study of cell functions is the core of research and development in biotechnology and provides the basis for processes such as gene splicing and genetic engineering. Comprehending the concepts outlined in the other chapters of this text will be much easier if the student has a basic knowledge of cell functions.

CHAPTER REVIEW

Student Learning Activities

1. Search the Internet and locate a research study dealing with cells. What potential application does the research have? Does it involve biotechnology? Are there ethical issues involved? If so, explain the issues. Share your findings with the class.

2. Using latex gloves, break open an egg into a dish. Locate the nucleus, the cytoplasm, and the cell membrane. Describe the function of each.

3. Draw a diagram of both an animal and a plant cell. Make a list of the differences between the two.

Fill in the Blanks

1. We can use _microbial_, _plant_, and _animal_ cells to provide useful _products_, processes and services.

2. _____ and _____ cells are similar in that they are both filled with _____, a gel-like substance that is about 80% _____ and contains the cell _____, proteins, the cytoskeleton, and other needed molecules.

3. Viral particles or _____ have some _____ to cells in that they contain _____ _____ and _____ _____.

4. Prokaryotic cells contain genetic material, but unlike _____ cells, this material is localized to the _____ _____ of the cell _____.

5. _____ can easily exchange or share genetic information through _____ _____ _____, where _____ or fragments of _____ may move between cells.

6. The functions of the cell membrane, or _____ membrane, are to enclose and _____ the cell's contents; _____ the movement of materials in and out of the cell; and allow _____ with other cells.

7. One of the most important _____ is the peanut-shaped _____, which functions to break down food _____ and supply the cell with _____.

8. Transporting proteins and other needed molecules throughout the cell is a team effort that involves the _____ membrane, the cell membrane, the _____ _____, the _____ _____, and vesicles that shuttle between these cell structures. These structures are all members of the eukaryotic _____ _____.

9. Early _____ cells divide and group together to form a ball-shaped mass called the _____, where the cells _____ and clump into a mass in a process called _____.

True or False . . . .

1. Scientists learned about cells through the processes of gene splicing and genetic engineering.

2. The invention of the electron microscope allowed scientists to see small cell parts and viruses for the first time.

3. Only eukaryotic cells contain genetic material.

4. Prokaryotes are the most numerous forms of life on Earth.

5. Bacterial cells such as *Escherichia coli* and *Lactobacillus* played an important role in early biotechnology.

6. All plants and animals are made up of prokaryotic cells.

7. The cell membrane is semipermeable, which means that it allows only certain materials to pass into and out of the cell.

8. The process of meiosis creates two identical daughter cells.

9. Interphase is not considered a true phase of mitosis because the cell is not actively dividing and is completing normal functions, such as growing and synthesizing materials.

10. A totipotent cell is a cell that is capable of developing into any kind of cell found in the adult animal's body.

Discussion . . . .

1. Explain how the electron microscope works and why it was so important for biotechnology.

2. Explain why viruses are not considered cells.

3. Compare and contrast eukaryotic cells and prokaryotic cells.

4. Discuss the differences between osmosis and diffusion.

5. Explain the function of chromatin, DNA, and genes.

6. Compare and contrast the structures of plant cells and animal cells.

7. List the differences between oogenesis and spermatogenesis.

8. List the phases of mitosis.

9. Describe the phases of mitosis.

10. Compare and contrast the three different kinds of stem cells.

CHAPTER 5

The Principles of Genetic Transfer

OBJECTIVES

When you have finished studying this chapter, you should be able to:

- Explain the function of DNA.
- Discuss nucleotide sequencing.
- Explain the role of genes in the transfer of characteristics.
- Discuss the concept of gene interaction.
- Define the process of transcription.
- Define the process of translation.
- Discuss the mapping of the human genome.

The Phenomena of Gene Transfer

One of the great mysteries of all times has been the ability of organisms to reproduce and create new organisms that resemble the parents. For thousands of years, humans have known that the offspring of plants and animals closely resemble their parents. Throughout most of history, they considered the transference of characteristics from parent to young as being part of the "miracle of birth" (Figure 5–1). Even though people did not understand how this process worked, they made use of the phenomena through selective breeding. They selected seed to plant from the plants that they thought were superior to other plants. Animals that best suited their purposes were saved for breeding, and throughout several generations of genetic transfer, superior animals were developed.

One phenomenon that people noticed was that although characteristics of parents were usually passed on to their offspring, offspring sometimes exhibited dramatically different characteristics. The first person to conduct systematic research into this puzzle was an Austrian monk named Gregor Mendel (Figure 5–2). Mendel lived in a monastery in the mid-1800s. He was an educated man who had studied science and mathematics and had a profound curiosity about the natural world. As mentioned in Chapter 2, he experimented with garden peas because of the differences in the appearances of the vines and fruit and decided that

Figure 5–1
Throughout most of history, people considered the transference of characteristics from parents to young as being part of the miracle of birth.

© iStockphoto//Maurice Van der Velden

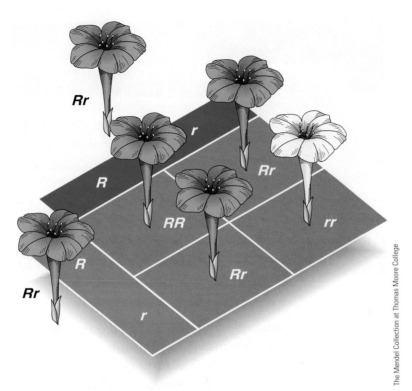

Figure 5–2
The first person to conduct systematic research in genetics was Gregor Mendel.

they would provide the basis of a study that would help determine how traits are passed on. As a result of the studies, Mendel developed the **law of segregation**. In essence, this law says that the factors, called **alleles**, responsible for the traits from each parent are separated and then combined with factors from the other parent at fertilization. Alleles are different forms of the same gene. In other words, each of the parents provides one of the two alleles for each particular trait. Mendel's experiments with the different traits led to the development of the **law of independent assortment**. This principle or law states that factors (genes) for certain characteristics are passed from parent to the next generation separate from other alleles that transmit other traits (Figure 5–3). This separation allows the tremendous amount of diversity among organisms. For example, an organism's color, size, shape, growth rate, and reproductive capacity are separate traits that can be passed to the next generation in any combination.

As advanced as Mendel's laws were at the time, no one understood how these characteristics were passed from generation to generation. Around the turn of the twentieth century, scientists found renewed interest in Mendel's laws

Figure 5-3

The law of independent assort-ment states that an organism's color, size, shape, and so on can be passed to the next generation in any combination. For example, a white cow can have a red calf.

of inheritance. Subsequent research revealed that genetic transfer takes place within the cells of organisms. This research occurred over about the first 25 years of the century, and simultaneously knowledge of the existence of **chromosomes** was developed. Once this knowledge was established, scientists then discovered the molecular basis of inheritance during the next 25 years. As mentioned in the previous chapter, the development of the electron microscope in the mid-1900s allowed scientists to thoroughly examine the inner workings of living cells. Around that time, two scientists, James Watson and Francis Crick of the Cavendish Laboratory in Cambridge, England, discovered how the building blocks of genetic transfer, known as **deoxyribonucleic acid (DNA)**, are organized in the cell and the process by which they replicate. Although these building blocks in the nucleus of cells had been discovered earlier, Watson and Crick's research on this molecule has led to a more complete understanding of just how traits are passed from parents to offspring.

Scientists in the latter half of the twentieth century uncovered the molecular mechanisms by which genetic information in cells is encoded and decoded (Figure 5–4). This allowed them to better understand the way the molecules of inheritance functioned in the cell by carrying information about physical characteristics and behavior from one generation to the next. This, in turn, provided the means to finally begin to comprehend how genetic transfer functions. Scientists still do not completely understand the

Courtesy of ARS/USDA

Figure 5–4
Location of specific DNA molecules was discovered in the latter part of the twentieth century.

process, but a tremendous amount of progress has been made in the knowledge we have. In the following paragraphs, our understanding of how the process works will be explained. If this process seems complicated, keep in mind that even with our advanced knowledge of genetic transfer, we probably understand less than 10 percent of the entire process.

DNA Sequencing

Higher-ordered animals and plants such as those produced in agriculture are composed of billions upon billions of cells. The DNA within the nucleus of each one of these cells contains a complete **genetic code** for creating an identical organism. DNA is composed of units called **nucleotides** that are made up of a sugar molecule, a phosphate molecule, and a nitrogen molecule containing chemicals called bases. The units or nucleotides of DNA are arranged together on a strand called a **helix** that resembles a long, twisted ladder. At each point on the helix, where the two halves of the ladder are connected, different nitrogen-containing bases— adenine (A), thymine (T), guanine (G), and cytosine (C)— are attached to each other at the center of the rung. These **nitrogen bases** are shaped so that each one can pair with only one particular base. Adenine (A) can pair only with thymine (T), and cytosine (C) can pair only with guanine (G) (Figure 5–5).

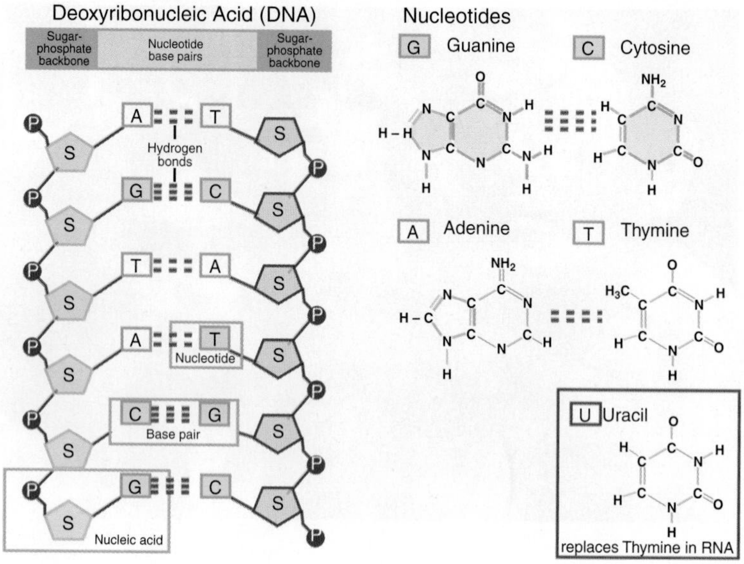

Deoxyribonucleic Acid (DNA)

Nucleotides

Courtesy of National Institutes of Health

Figure 5–5

DNA is composed of a sequence of nucleotides linked by nitrogen bases.

Genes

Each side of the twisted ladder or helix is composed of a specific sequence of nucleotides that have a nitrogen base at the center. A nucleotide containing adenine may be connected to a nucleotide containing cytosine, or one containing guanine, or another nucleotide containing adenine. It is the sequence of DNA's nucleotides that functions as the genetic code. Nucleotides are arranged in functional segments called **genes** that encode specific biological traits. This is what controls the creation of certain characteristics in the organism. Within each organism are sequences of different sizes. One sequence may be relatively short and control a certain trait. Another sequence may be relatively long and control a different trait. For example, as far as we know now, the shortest sequence in the human genetic code is about 50 nucleotides, and the longest contains about 250 million.

Individual genes may work independently to govern biological traits. For example, one gene may control hair color in an animal, and another gene may control the hair length or the texture of the animal's hair. Genes control the color of flowers, the amount of protein or carbohydrates in a seed, and the height of the plant at maturity. As mentioned earlier, different forms of the same gene are called alleles.

Courtesy of ARS/USDA

Figure 5–6
A combination of environmental and genetic factors control how an animal looks.

Environmental factors may also play a crucial role in the development of biological traits (Figure 5–6). The quality and quantity of nutrition can control the size of a plant or animal, and the climate may have an effect on everything from reproduction to how long the organism lives. However, no matter what the environmental factors are that surround an organism, the characteristics of a plant or animal are limited to the genes possessed by the organism. For example, a sow may have genetic trait to bear and nourish a litter of 10 piglets, but if she does not receive the proper nutrition or the climatic conditions are harsh, she may never be able to have as many as 10 piglets. On the other hand, if a sow is genetically capable of bearing and nourishing only six piglets, that is all she will ever have even though she may have ideal nutrition and perfect climatic conditions.

The expression of the genetic makeup of an organism, including all of its variant genes or alleles, is called the **geno-type**. When an organism reproduces or a cell divides, it is the genotype that is passed to the next generation. The observed biological characteristics brought about by the environment and the genetic code is called the phenotype. In other words, the **phenotype** is how the organism looks as a result of a combination of the expression of its genes and the effects of the environment in which it lives. This may be expressed by the formula

Genotype + Environment = Phenotype

In plants and animals that reproduce sexually, every gene that comes from the male is paired with a gene of the same type from the female. For example, the gene that controls the color of a flower is made up of a pair of "color" genes; one half of the pair (or one allele) comes from the father, and one half of the pair (or one allele) comes from the mother. As previously mentioned, a pair of genes that controls a specific characteristic is called an allele. If both of the alleles are the same form, the genes are said to be **homozygous**, and the flower will be the color called for by the genes. For example, if both are coded for a red flower, the flower will be red, or if both are coded for a white flower, the flower will be white.

However, genes from a plant with a white flower may unite with genes from a plant with a red flower. In this case, the alleles are different forms and are said to be **heterozygous**, and the color of the offspring's flower will be determined by the dominant allele. This means that the effect of one gene will take priority over the effect of the other gene. If the white flower is dominant, the flowers from the plant offspring will be white. If the red color is dominant, the flowers will be red although the plant contains genes for both white and red flowers. The resulting genes are said to be heterozygous. If the red flower is dominant and is crossed with another plant with only white flower genes (all genes homozygous), half the genes that control color will be for red flowers, and half will be for white flowers. The resulting flowers will all be red. If the offspring of these plants are then crossed with a plant having genes for white flowers, at least some of the next generation of plants will have white flowers.

Punnett Square

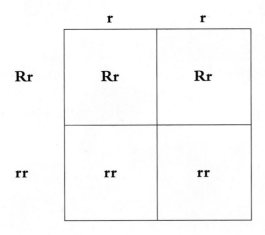

Figure 5–7
A Punnett square provides a scheme that represents the likelihood of characteristics of offspring from heterozygous parents.

Figure 5–7 is called a Punnett square and illustrates how genetic dominance can work with heterozygous flowers, where *R* represents the dominant red color and *r* represents the recessive white color. In this case, the white flower is represented with an *rr* across the top, and the red flower is represented on the side of the chart as *Rr*. Half of the flowers will be red, and half will be white.

However, there are exceptions to the **rule of dominance**. Some pairs of genes may not have a gene that is dominant over the other. They are of equal power and are said to be **codominant**. For example, neither red flowers nor white flowers may be dominant but may be codominant. This means that neither gene can override the effect of the other. When this happens, a blending of the colors will likely occur, and the flowers may be pink.

A very good example of codominance may be found in shorthorn cattle. Purebred shorthorns may be red, white, or roan (Figure 5–8). Cattle that are completely red carry genes that call for red color only (*RR*), cattle that are completely white carry genes for white color only (*WW*), and cattle that are roan or spotted carry a gene for red and one for white (*RW*). In this case, neither the red (*R*) nor the white (*W*) gene is dominant. The color of the animal will be a combination of red and white and will be spotted, or roan, in color. These cattle can have a variety of color combinations and still be registered as purebred shorthorn cattle.

Courtesy of ARS/USDA #K2610-10

Figure 5–8
A roan-colored shorthorn may have a parent that is solid red and a parent that is solid white.

The coat color trait in cattle is one good example to help understand how pairs of alleles or genes might interact to produce a biological trait. The same general principles can be applied to other traits, such as horned or polled, tall or short, and so on. However, the entire process of defining the characteristics of the animal by the genetic makeup is tremendously more complicated. For instance, genes that are not alleles (matched pairs that control a characteristic) may interact to cause an expression that is different from the coding on the genes. This interaction is called *epistasis*.

Another factor in genetic transfer of characteristics is that of the additive expression of genes. This means that a number of different genes may be "added" together to produce a certain trait in an animal. For instance, the amount of milk produced by the female is controlled not by a single pair of alleles but by several pairs. The size and body capacity of the female, the ability to produce the proper amounts of hormones, and the mammary gland's size and functioning ability are all controlled by different pairs of alleles (Figure 5–9). Yet all of these factors contribute to the female's overall ability to produce milk. The same thing may be said about the animal's rate of gain or its ability to efficiently reproduce. Several genetically controlled factors, such as body size and structure, can affect the animal's ability to grow rapidly and efficiently. A heifer's pelvic size, shape of the genital tract, and output of sex hormones are all controlled by different genes and are all factors in reproduction efficiency.

Figure 5–9
The milking ability of a female may be controlled by several genetic factors.

Chromosomes

Genes are grouped together on dense physical structures known as **chromosomes**. Chromosomes are composed of chromatin, which is made up of DNA tightly wound around structural proteins, and are housed within the nucleus of each eukaryotic cell (Figure 5–10). Remember that eukaryotic cells have a nucleus and prokaryotic cells have no true nucleus. Chromosomes are also present in prokaryotic cells, though prokaryotic chromosomes are typically small, circular structures with a simple architecture. The biological purpose of chromosomes is to wind up and organize the DNA in a structure that is easy to duplicate and divide during cell division (mitosis). Chromosomes line up in the center of the cell and are carefully separated to make sure that each daughter cell receives a full copy of genetic information.

Careful and accurate division of genetic information is also critical in the specialized cell division that forms gametes (meiosis). Remember from Chapter 4 that sets of chromosomes are divided in the process of meiosis, which produces the sperm or egg. When the sperm and egg are united at conception, the chromosomes are paired together and form new sets of chromosomes for the new organism. Figure 5–11 illustrates the number of chromosomes that can be found in the cells of various animals. When gametes unite during sexual reproduction, two complete sets of chromosomes are brought together in the new individual. The nucleus of

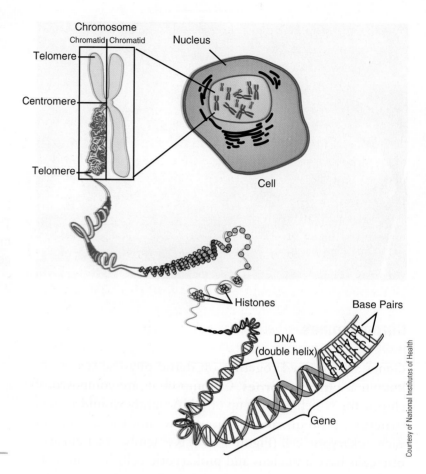

Figure 5–10
Genes are grouped together on dense bodies known as chromosomes within the nucleus of a eukaryotic organism's body.

Courtesy of National Institutes of Health

CHARACTERISTIC NUMBERS OF CHROMOSOMES IN SELECTED ANIMALS

Animals	Chromosome Number (2n)
Donkey	62
Horse	64
Mule	63
Swine	38
Sheep	54
Cattle	60
Human	46
Dog	78
Domestic cat	38
Chicken	78

© Cengage Learning 2013

Figure 5–11
Some example of the number (2n) of chromosomes of various animals.

sexually reproducing plants and animals contain two sets of chromosomes—one set from each parent.

Chromosomes may be classified as either autosomes or sex chromosomes. Autosomes are chromosomes that carry genes for all the traits of an organism except sex determination. Sex chromosomes are those that carry genes that will determine the gender (male or female) of an organism. In many higher animals, including humans, there are two types of sex chromosomes, designated "X" and "Y." The X chromosome is large and contains many genes. The Y chromosome is much smaller and includes genes for "maleness." In animals with this system of sex determination, females have two X chromosomes, written as "XX." Males have two different sex chromosomes, written as "XY" (Figures 5–12A and 5–12B). When the gametes (sperm or egg) are formed during meiosis, females have only X chromosomes to donate to their eggs. Males, on the other hand, may pass an X or a Y chromosome on to their sperm. When egg and sperm unite at conception, the presence of an XX or XY pair of chromosomes determines whether the resulting animal will be male or female.

RNA and the Process of Transcription and Translation

When cells divide, the genetic material in the parent cell must be passed along to the new cell. As previously mentioned, this material is contained in a genetic code held in the chromosomal DNA of all cells. The basic process of decoding genetic information is similar in prokaryotic and eukaryotic cells. This genetic code is copied using **ribonucleic acid (RNA)**, a substance in the living cells of all organisms that carries genetic information needed to form protein in the cells. Within the nucleus of the parent cell, the code of genetic information is copied, or transcribed, to a molecule of a type of RNA called **messenger RNA (mRNA)** in a procedure that is somewhat like the replication of DNA in mitosis. This process is known as **transcription**. Once mRNA copies the information, it leaves the nucleus and moves to the cytoplasm of the cell, where the code specifies the particular **amino acid** that is to be used to make up a protein (Figure 5–13).

Within the cytoplasm, the genetic information is translated into the amino acid by a different type of RNA called **transfer**

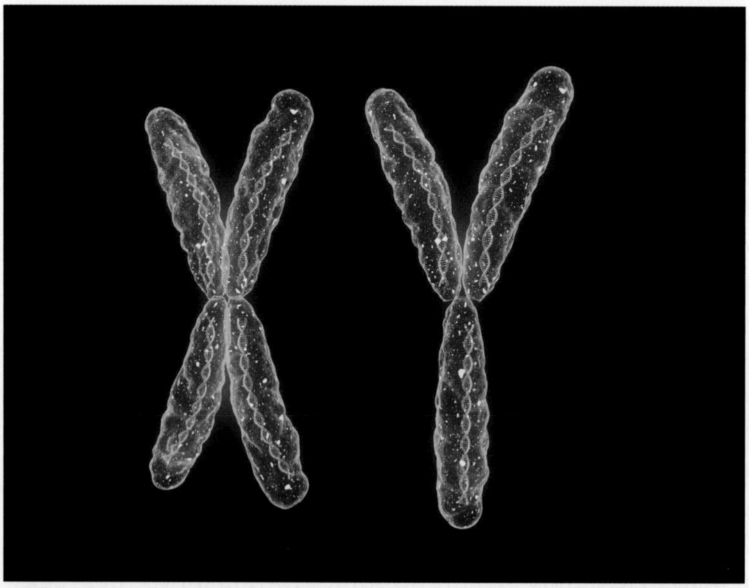

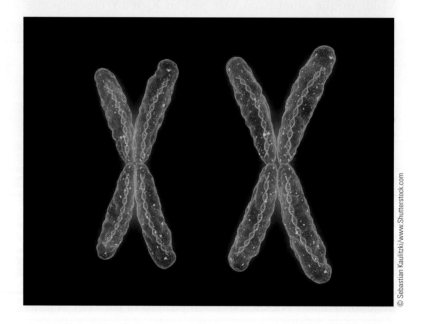

Figure 5–12A
Males have an X and a Y chromosome.

Figure 5–12B
Females have 2 X chromosomes.

RNA (tRNA). This process, known as translation, decodes the information and makes it available for use. Within the cell, this information is carried to organelles called ribosomes (see Chapter 4), where another type of specialized RNA, known as **ribosomal RNA (rRNA)**, synthesizes amino acids.

Amino acids are the building blocks of proteins, and most organisms are composed largely of protein. Most higher-order animals are made up of about 100,000 types of protein. The genetic code for each amino acid is contained in a series

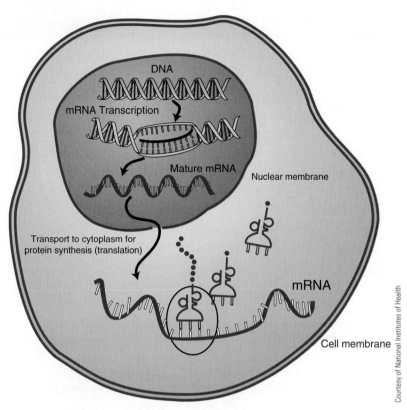

DNA

mRNA Transcription

Mature mRNA

Nuclear membrane

Transport to cytoplasm for
protein synthesis (translation)

mRNA

Cell membrane

Courtesy of National Institutes of Health

Figure 5–13
The transfer of DNA is
accomplished by several types
of ribonucleic acid (RNA).

of three nucleotides, called a **codon**, that tells the cell to
build a certain type of amino acid (Figure 5–14). A gene may
have over 1,000 bases or nucleotides, or it may be quite a bit
shorter, depending on the characteristic the codon specifies.
However, only about 10 percent of the genome is known to
have codons that transcribe amino acids.

These codon sequences are called **exons**. Intermingled
throughout the genes are sequences called **intron sequences**,
which contains no codes (Figure 5–15). There are other regions of the genome that have no coding function, and the
purpose of these regions is not fully understood by scientists.

Genome Sequencing (Gene Mapping)

The series and sequences of nucleotides arranged on chromosomes is the genetic code that controls all the characteristics of
an organism that are passed from one generation to the next.
A goal of scientists during the 1980s and 1990s was to figure
out just how the nucleotide sequence of DNA for various organisms is arranged in all of the genes and chromosomes of

Using Genetics to Build a Better Tomato

Agricultural Research Service (ARS) researchers are working to save our tomatoes—or at least some of them.

Tomatoes spend so much time on shelves and in refrigerators that an estimated 20 percent are lost to spoilage, according to the USDA Economic Research Service. Autar Mattoo, an ARS plant physiologist with the Sustainable Agricultural Systems Laboratory in Beltsville, Maryland, is trying to change that. Mattoo is working with Avtar Handa, a professor of horticulture at Purdue University, to enhance tomatoes so that they offer not only better taste and higher nutrient levels, but also a longer shelf life.

Mattoo, Handa, and Savithri Nambeesan, a graduate student working with Handa, recently focused on manipulating a class of nitrogen-based organic compounds known as "polyamines" that act as signals and play a role in the plant's growth, flowering, fruit development, ripening, and other functions. Polyamines have also been linked to the production of lycopene and other nutrients that lower our risk of developing certain cancers and other diseases, making them a prime target for investigation, according to Mattoo.

"We wanted to see if we could increase the levels of polyamines in tomatoes and then investigate their biological effects," Mattoo says.

The researchers introduced a yeast gene, known as "spermidine synthase," into tomato plants specifically to increase production of a single polyamine—spermidine. Spermidine is found in all biological organisms and is one of three polyamines believed to modulate the plant-ripening process.

In a Beltsville, Maryland, greenhouse, plant physiologist Autar Mattoo (center) points out features of a genetically improved tomato line to postdoctoral fellow Vijaya Shukla (left) and biological technician Joseph Sherren.

The results showed that introducing the gene not only increased spermidine levels and vegetative growth, but also significantly extended the tomato's postharvest shelf life. Shriveling was delayed by up to 3 weeks, and there was a slower rate of decay caused by tomato plant diseases. The tomatoes also had higher levels of the antioxidant lycopene. The study shows for the first time that spermidine has its own effects, independent of the other polyamines, extending shelf life and increasing growth. Polyamines are found in other plants, so the work could assist in efforts designed to extend the postharvest shelf life of other crops.

"We know that in tomato fruit, the signaling machinery continues to function late into the ripening process. By designing genes that would lead to higher levels of polyamines, it should be possible to modulate ripening and influence nutrient levels as well," says Mattoo.

The use of molecular genetics to enhance tomatoes has faced some resistance from consumers and industry. But scientists have used such molecular techniques for years to develop improved varieties of corn, soybeans, and cotton, and Mattoo is confident that in time the approach will become more widely accepted as its benefits are better understood.

—By Dennis O'Brien, Agricultural Research Service information staff.

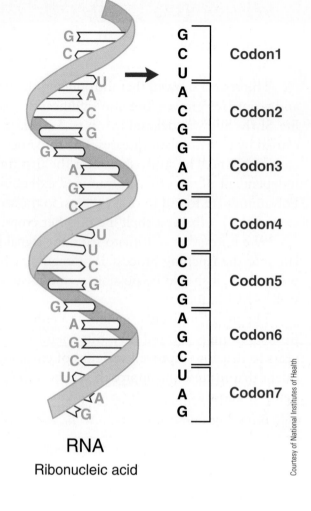

RNA

Ribonucleic acid

Codon1: G C U
Codon2: A C G
Codon3: G A G
Codon4: C U U
Codon5: C G G
Codon6: A G C
Codon7: U A G

Figure 5–14

The genetic code for each amino acid is contained in a series of three nucleotides called a codon that tells the cell to build a certain type of amino acid.

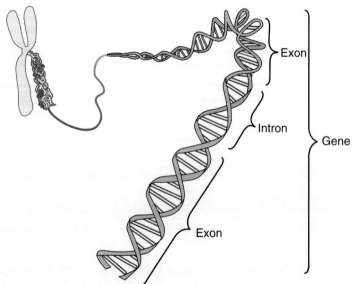

Exon

Intron

Gene

Exon

Figure 5–15

A gene contains areas of coding called exons and areas of noncoding called introns.

the entire body of an organism. The reasoning behind the goal was that once we understood the arrangement of the nucleotides on the chromosomes, we could then begin unlocking the mechanisms that control certain traits. Implications for this knowledge are almost beyond our comprehension. For example, if we know the precise gene that controls any number of genetically transmitted diseases in humans, the specific gene may be treated, and the disease may be eradicated. In fact, the control genes for several diseases, such as Parkinson's disease, have already been identified—although it took scientists over 7 years to locate this specific gene (Figure 5–16).

DNA sequencing technologies are getting faster and less expensive all the time. We can now sequence billions of nucleotides in a matter of days rather than years, opening up many possibilities for genotyping plants and animals.

The genome of humans, as well as hundreds of other species, is made available to the public on the National Center for Biotechnology Information (NCBI) website. Scientists using public monies to fund their research are required to publish all DNA and protein sequences to the databases maintained at NCBI. However, although we have a mountain of DNA sequence data for humans and many other organisms, big questions remain to be answered. For each gene in the database, numerous laboratory experiments must be carried out to characterize the biological function of the gene. In addition, as discussed in this chapter, genes act in concert to yield certain traits. We have yet to understand how gene interactions are occurring within cells and

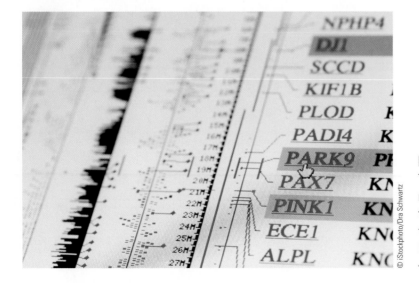

Figure 5–16
The human genome has been mapped and specific genes identified. Here the gene (PARK9) for Parkinson's disease has been identified.

in response to environmental signals. The ultimate goal is to characterize the functions and interactions of all genes in all biological systems. Deep knowledge of our own genetic code may unlock the secrets of genetically transmitted disorders and lead to cures for many other types of diseases.

The goal of completing the map of the human genome—that is, the complete DNA sequencing of the human—was completed in 2001 as the result of a gigantic project that focused on locating the genetic sequences (Figure 5–17). Other more simple organisms, such as bacteria, had been mapped earlier, but no organism as complicated as a human being had ever been mapped before. The human genome was chosen for the project because of the tremendous potential that could be realized by locating all of the human genes.

It appears that humans possess about 34,000 genes. While this may seem like a large number, up until about a year before the completion of the project, scientists predicted that the human gene map would contain up to 150,000 genes. The smaller number came as a surprise, considering that lower-order worms may contain half as many genes as those of a human. Another surprise was that humans share around 200 genes with bacteria. The mystery now is how an organism as complicated as a human being could have only twice as many genes as that of a worm as well as sharing common genes with bacteria. The best theory is that human genes are

Figure 5–17
The complete DNA sequencing of the human genome was completed in 2001.

© TEK IMAGE/SPL/Science Photo Library/Getty Images

more complex, producing more types of protein for translation than the genes found in the lower-ordered organisms.

Myths surrounding perceptions about racial differences were dispelled by the discovery that all individual humans have 99.9 percent of the same, identical genes, and, in fact, there are more differences *among* races than *between* races (Figure 5–18). The question is raised that if we are this close in genetic makeup, why are individual humans so different? The answer is that, like all genomes, the human genome is composed of the sequencing of letters representing the organization of different nitrogen-containing bases, adenine (A), thymine (T), guanine (G), and cytosine (C), attached to each other at the center of the chromosome. For the entire human genome, there are over 3 billion letters. This amounts to volumes of letters, the equivalent of over 200 New York City phone books. If humans have 99.9 percent the same genetic makeup, this means that .1 percent is different. If we calculate .1 percent of 3 billion, there are 3 million letter codes different in humans—and this accounts for a lot of variability between individuals.

Another puzzle is that there are apparently large areas of coding (intron sequencing) on the chromosomes that create very few genes. This is the reason that scientists were surprised to find that the 3 billion letter code created only about 34,000 genes rather than the expected 100,000-150,000 genes. When the project to map the human genome began, some scientists considered it a waste of time

Figure 5–18
Genetic research has shown that there are more differences among races than between them.

to code the areas that do not code for protein. However, as new insights are gained, it is becoming clearer that although these areas do not code for proteins, they play essential roles in the genetic transfer process. For example, some of the sequences are **controller sequences** that actually turn on or turn off a particular gene. These controllers may tell a cell when to stop multiplying or may control the release of enzymes as well as many other types of control mechanisms.

Even though scientists have mapped the entire genome for the human body and many other higher-ordered organisms are currently being mapped, the amount of information that can be put to use is still limited. A genome map simply tells us where the sequencing of nucleotides takes place on the chromosomes. We are still a long way from being able to determine how this sequencing controls for specific genes and the characteristics controlled by the genes. When we are able to locate the particular gene that controls each different characteristic of the human body, we will then be able unlock the secrets of genetically transmitted disorders as well as to cure many other types of disease (Figure 5–19). This knowledge will naturally bring about a lot of controversy over the manipulation of genes, a topic to be covered in more detail in a latter chapter.

Summary

Humans have always wondered at the marvel of plant and animal reproduction that creates offspring resembling the parents. The process of understanding how the transfer of characteristics from one generation to the next has slowly evolved over the past 100 years. Only within the past 20 years we have begun to understand how the process works. From Mendel in the 1800s to now, we have a good understanding of how genes are inherited from parent to offspring. The current challenge is to learn more about the function of specific genes, including when and where they are transcribed and translated in cells. An important modern tool for characterizing and understanding genes is genome sequencing. Even though we have gained a tremendous amount of knowledge about gene transfer, most of

Figure 5–19
When we are able to locate the particular gene that controls each of the different characteristics of the human body, we will be able to unlock the secrets of genetically transmitted diseases and disorders.

© Ryan McVay/Photodisc/Getty Images

the knowledge about the process remains to be discovered. As soon as scientists unravel one mystery, another emerges. The next few years will prove to be an interesting and exciting time as scientists solve the great perplexities of gene transfer. Exciting advances in medicine, agriculture, and basic biology rest on our understanding of living systems and the complexity of the genetic code.

CHAPTER REVIEW

Student Learning Activities ....

1. Conduct an Internet search to locate information about other genomes that have been mapped. Choose one of the projects and report to the class. How much of the genome has been mapped? What surprises were found? What are the potential uses for the genome?

2. Research a human disease that is hereditary. Compile a report on genetic research that promises to cure the disease. Share the report with the class.

3. Choose a breed of livestock and write a report on how the breed was developed. What was the root stock? What characteristics were sought?

4. Choose a breed of livestock or other animal and research the breed. What problems are associated with the particular breed? Explain how gene splicing might help alleviate the problem.

Fill in the Blanks ....

1. Two scientists, _____ and _____, discovered how the building blocks of genetic transfer, known as _____, or _____ _____, are organized in the cell and the process by which they replicate.

2. DNA is composed of units called _____ that are made up of a _____ molecule, a _____ molecule, and a _____ molecule containing chemicals called _____.

3. _____ traits are an exception to the rule of _____ and result in a phenotype that is a blending of traits.

4. Genes that are not _____ may interact to cause an expression that is _____ from the coding on the _____. This interaction is called _____.

5. When the _____ and _____ are united at _____, the chromosomes are _____ together and form new sets of chromosomes for the new organism.

6. The genetic code is copied using _____ _____ (RNA), a substance in the living cells of all organisms that carries _____ _____ needed to form _____ in the cells.

7. _____ _____ are the building blocks of _____, and most organisms are largely _____ of protein.

8. The genetic _____ for each amino acid is contained in a series of three _____, called a _____, that tells the cell to _____ a certain type of amino acid.

9. Originally, scientists thought the human genome contained over _____ genes, but while mapping the human genome, they found that humans contain only _____ genes. About _____ of these genes are shared with bacteria.

10. Some areas of the genome that do not code for _____ contain sequences called _____ _____ that actually turn on or turn off a particular _____.

True or False

1. Adenine can pair only with cytosine, and thymine can pair only with guanine.

2. Genes are always made of the same number of nucleotides.

3. If a dominant allele is present in an organism, the organism will display the dominant phenotype.

4. Autosomes are the chromosomes that carry the genes for sex determination.

5. Genetic information is transcribed into messenger RNA in the nucleus.

6. Transfer RNA (tRNA) synthesizes amino acids.

7. Cell organelles called ribosomes are necessary for translation to occur.

8. DNA sequencing technologies have not progressed much over the past 20 years.

9. Once the genetic sequence of an organism has been mapped, only scientists may access and use the information.

10. There are more genetic differences among humans of the same racial group than there are between humans of different racial groups.

Discussion

1. Explain Mendel's law of segregation and law of independent assortment.

2. Distinguish between the genotype and phenotype of an organism.

3. Explain how two organisms can be genetically identical yet still have physical differences.

4. Explain the difference between heterozygous and homozygous genes.

5. How is sex determined in higher animals?

6. Identify the products of transcription and translation.

7. Name the three types of RNA and describe their function.

8. Distinguish between exons and introns.

9. Why do scientists think it is important to map the genomes of organisms?

10. Discuss the limitations of gene mapping.

CHAPTER 6

Producing Genetically Modified Organisms

OBJECTIVES

When you have finished studying this chapter, you should be able to:

- Define genetically modified organism.
- Explain how bacteria are used to produce genetically engineered products.
- Discuss the steps involved in placing foreign DNA into an organism.
- Describe how a segment of DNA is selected and removed from a DNA strand.
- Explain how vectors are used to insert DNA into an organism.
- Describe the process of splicing DNA into plants.
- Describe the process of splicing DNA into animals.
- Explain how transgenic animals are used in genetic research.

Introduction

In the history of science, there have been few developments that have generated more excitement than the ability to genetically engineer living organisms. Early producers used selective breeding to choose the plants and animals with the most useful traits as parents for subsequent generations. The process of **selective breeding** has allowed humans to slowly change the genetic makeup of domesticated species over thousands of years. Think of the diverse and numerous varieties of plants and breeds of that share common wild ancestry. It took thousands of years of selective breeding for specific observed traits to develop domestic varieties and breeds. At first, this process was mostly by trial and error, but in later years the process has been refined into a more exact science. Pollination control and isolation in the breeding of plants has dramatically increased the genetic improvement of plants (Figure 6–1). Unfortunately, some traits with negative health consequences accompanied the selected positive traits for a number of dog breeds.

In contrast to traditional selective breeding, **genetic engineering**, or **gene splicing**, allows for the very specific isolation and movement of genes for desirable traits without the inclusion of less desirable traits. Through the addition, deletion, or modification of genes for useful traits, we are now able to create genetically modified organisms (GMOs) that help maximize their benefit to humans.

In addition to genetic modification, other modern techniques of plant and animal breeding have been developed and play important roles. In animals, the advent of methods such as artificial insemination and embryo transfer, coupled with computerized progeny data, have allowed great strides in animal improvement. Although it is by no means the only way to change the genetic structure of organisms, the term **genetically modified organisms** has come to refer to those organisms that have had genetic material removed and/or inserted in order to change a particular trait or traits of the organism. These plants or animals are referred to as transformed plants or animals, and the process is known as gene splicing, or genetic engineering.

No amount of selective breeding can achieve the rapid results gained by actually removing and inserting genetic materials into living organisms (Figure 6–2). Not only that,

Figure 6–1

Conventional methods, such as pollination control, greatly increased the genetic improvement of plants.

USDA/ARS. Photo by Stephen Ausmus

© Fuse/Getty Images

Figure 6–2
No amount of selective breeding can achieve the rapid results gained by actually removing and inserting materials into living organisms.

but if a desirable trait does not exist in a particular species, it may be difficult or impossible to breed it in through traditional means.

In the late twentieth and early twenty-first centuries, this process created more enthusiasm and apprehension than almost any other breakthrough dealing with plants and animals. Theoretically, the potential exists to be able to remove any unwanted characteristic and to add any beneficial characteristic to the plants and animals produced in the agricultural industry. In fact, the use of GMOs is widespread throughout agriculture, medicine, and other industries. Specific uses and advantages of genetically modified plants and animals are discussed in detail in other chapters.

The molecular tools and techniques that allow manipulation of DNA sequences in the laboratory are collectively referred to as recombinant DNA technologies. DNA combined from different sources is called **recombinant DNA**. Organisms produced by recombinant DNA technologies to transplant genetic materials between different types of organisms are called GMOs, or **transgenic organisms**. The genes of interest that are moved into an organism using recombinant DNA technology are also known as **transgenes**. For example, transgenes from bacteria have been transferred to agriculturally important plants, such as corn and cotton, in order to make these crops less susceptible to insect damage (Figure 6–3). Genes for the production of human growth hormones have been successfully transferred

Courtesy of ARS/USDA

Figure 6–3
Genes from bacteria have been successfully inserted into crop plants to help them resist insect damage such as this.

into the genomes of mice and other animals, doubling their growth. Genes for many different types of useful proteins, such as therapeutic drugs, vaccines, and industrial enzymes, have been genetically engineered into bacteria and fungi used in biomanufacturing. Human genes can be transplanted into pigs with the hope of producing tissues for use in human bodies. Although this technology is still experimental, the possibilities for these types of organisms are almost limitless.

Genetically Modified Microorganisms

Not only was the first genetic splicing done using bacteria, but the first practical use for transgenic organisms involved bacteria. Recall from an earlier chapter that these one-cell organisms have circular-shaped pieces of DNA called **plasmids** that float freely in the cell's fluid. **Enzymes** are protein substances that create or speed up chemical processes in the body of an organism. Enzymes play a vital role in genetic engineering. By selecting and using the proper enzyme, scientists cut out part of the plasmid DNA and insert DNA from another organism. The DNA replicates, and the new bacteria produced from the spliced DNA holds the desired characteristics. One of the first practical uses of genetic engineering was the biomanufacturing of human insulin. Insulin is a hormone produced by the human pancreas that helps to regulate the metabolism of carbohydrates. A condition known as diabetes results if the body does not manufacture enough insulin. Millions of people have this condition that causes severe health problems and can lead to early death if not treated properly.

The treatment for diabetes is to give patients daily doses of insulin to make up for the lack of insulin that ordinarily would be produced by their own bodies (Figure 6–4). When insulin was first used as a medicine to treat diabetes, it was harvested from the pancreas of cattle or pigs that had been slaughtered. Since only a very tiny amount could be obtained from an animal, the process was expensive. In addition, animal-derived insulin was not exactly the same as human insulin, and it triggered allergic reactions in some patients.

With the development of the technology involved with gene splicing, the problems created by the use of animal insulin were solved. Scientists were able to isolate the DNA sequence that regulates the production of insulin in the human body. Once this segment was found, scientists developed a technique for splicing this segment into the DNA of the *Escherichia coli* bacteria (Figure 6–5). This bacteria was chosen because it is so common and actually resides in the colon of all humans. Scientists understood the biology, genetics, and growth requirements of *E. coli*, and there were nonpathogenic lab strains available for use. It was a logical choice for the first biomanufacturing organism. In this technique, once the human DNA is extracted and spliced into the bacteria, all bacteria reproduced from the genetically altered bacteria have the insulin manufacturing genes. Since the genetically altered bacteria are weaker than the ordinary bacteria, they are grown under carefully controlled conditions. In a process called fermentation, the bacteria are produced in enormous quantities. Given the proper nutrients and environmental conditions, the bacteria grow rapidly and produce the protein of interest (insulin), acting as small biological factories.

Trillions of the genetically altered bacteria are reproduced until the proper amount are produced. Then they are removed from the fermentation tanks and taken apart to retrieve the insulin produced by the bacteria. The material is then separated and purified, and the remains of the bacteria are destroyed. By keeping a supply of the genetically altered bacteria, a relatively inexpensive, ready supply of insulin is available for those people who need it.

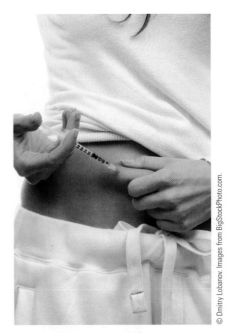

Figure 6–4
Many people have to inject themselves with insulin to treat a condition called diabetes.

Human Insulin Production

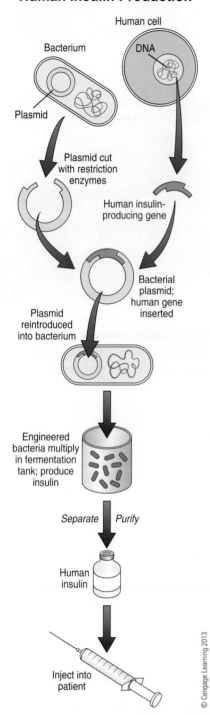

Human cell

Bacterium

DNA

Plasmid

Plasmid cut with restriction enzymes

Human insulin-producing gene

Bacterial plasmid; human gene inserted

Plasmid reintroduced into bacterium

Engineered bacteria multiply in fermentation tank; produce insulin

Separate   *Purify*

Human insulin

Inject into patient

© Cengage Learning 2013

Figure 6–5
Human insulin is produced by genetically altered bacteria.

Once the technology for producing insulin was perfected, bacteria became the manufacturing centers for many substances used to make the lives of humans better and more productive. Vaccines that are used to impart immunity into animals and humans are produced in much the same way as insulin. Also hormones that control growth and other bodily functions are produced using genetically engineered bacteria.

Another example of genetically modified bacteria being put to use is the substance called **bovine somatotropin (BST)**. BST is a hormone composed of protein produced in the pituitary gland of cattle. This hormone helps control the production of milk by assisting in the regulation of nutrients in the production of milk or fat. Certain breeds of cattle, such as Holstein, Jersey, and Guernsey, are known as dairy breeds and produce more milk and less fat than breeds such as Hereford, Angus, and Charolais that are grown as beef animals (Figure 6–6). One of the reasons is that the dairy breeds produce more BST. If these cows are given a supplementary dose of BST, the cows produce less fat and more milk. Scientists knew this for many years, but the cost of artificially producing BST prevented commercial use. Through genetic modification and fermentation of *E. coli* bacteria, the hormone is now produced at a relatively low cost.

Insulin and BST are two early examples of biomanufacturing using genetically modified microbes. It is now very common for pharmaceutical and industrial chemical companies to produce many types of useful proteins through biomanufacturing processes, including vaccines, hormones, and industrial enzymes. In addition to bacteria, many different types of cell cultures may be used in biomanufacturing to produce proteins. Depending on the protein that needs to be produced, biomanufacturers may choose to use genetically modified bacterial, fungal, insect, animal, or plant cell cultures. In the next section, the basics of genetic engineering will be discussed along with a look at the process of creating genetically modified plants and animals.

The Process of Genetic Engineering

Genetically modifying prokaryotes and other organisms that reproduce asexually, such as budding yeast, is not as difficult as attempting to genetically engineer a higher eukaryote.

Figure 6–6
Genetically altered bacteria is used to produce a hormone called somatotropin that helps increase milk output in dairy cows.

Figure 6–7
Scientists spend a lot of time finding and characterizing genes. This scientist is locating a gene in the soybean genome.

Plants and animals have a lot more genetic information to characterize, and their thousands of genes are under tight regulatory controls. Strict regulation of which genes are "on" and "off" allow plants and animals to grow and develop in complex ways. So, simply inserting a new gene into a plant or animal genome does not always guarantee a predicted downstream biological result. Scientists must spend a lot of time finding and characterizing genes and gene regulatory DNA segments, called **promoters** (Figure 6–7). Promoters control the expression of genes by interacting with the transcriptional machinery of the cell. These interactions must go smoothly in order to get the gene expression process "working" in genetically modified plants and animals.

Genetic engineering in plants and animals is a long and labor-intensive process that has been helped by the wealth of genomic data now available.

In order to genetically engineer bacteria, plants, or animals, there are a few basic technical questions to consider. First, which DNA segment(s) or gene(s) of interest will be engineered into the organism (gene discovery)? Second, how will these DNA segments be isolated, copied, and delivered to target cells (recombining DNA and cell transformation)? Third, how will we be able to tell if genetic engineering/delivery and integration of DNA segments has been successful (selecting transformants)? Ultimately, which physical traits or behaviors result or change in the GMO compared to the wild type, and are we able to measure these changes (phenotypic assays)?

Gene Discovery

To identify genes of interest, scientists rely on two basic genetic screening approaches: forward genetics and reverse genetics. In **forward genetics**, scientists observe the physical characteristics to find organisms with interesting or useful characteristics (Figure 6–8). This is called phenotypic screening. Natural populations of the organisms may be examined, but to increase the chance of finding an unusual trait, a mutation population is often generated using radiation or chemical mutagenesis. Remember that a mutation is an accident or an abrupt change in heredity. Mutation breeding of crop plants is essentially a forward genetics approach. Once

Figure 6–8

In forward genetics, scientists observe the physical characteristics to find organisms with useful characteristics.

© Auremar/www.Shutterstock.com

interesting mutant characteristics are identified, scientists use a number of molecular and biochemical approaches to figure out which genes are responsible for the interesting trait(s).

In **reverse genetics**, scientists have some knowledge, perhaps gained from a genome sequencing (gene mapping) project, which leads them to investigate the biological function of a specific gene. They may use a number of different recombinant DNA technologies to knock out or disrupt the gene of interest and then screen the resulting modified organisms for changes in the physical characteristics or phenotype. Later in the chapter, methods for creating gene knockout in animals will be discussed.

Recombining DNA

In order to genetically engineer cells, scientists need a way to "cut and paste" specific pieces of DNA. Fortunately, molecular tools that do just that exist in nature. A group of enzymes, called the **restriction endonucleases**, recognize specific DNA sequences and cut the DNA into pieces. Enzymes are proteins that catalyze or speed up chemical reactions in cells. This group of useful enzymes was discovered in bacteria in 1970. Their function is to serve as a bacterial defense system against viruses that try to invade. They do this by chopping the virus DNA into pieces. The discovery and purification of restriction enzymes made genetic engineering possible.

Another challenge is the storage, copying, and delivery of recombinant DNA segments into cells. Recall from an earlier chapter that prokaryotes often contain circular, self-replicating pieces of DNA called plasmids. In nature, plasmids sometimes move between bacteria and carry genes for useful traits, such as the ability to form resistance to antibiotics. Scientists have adapted plasmids for use as "gene delivery" vehicles, or **vectors**, that carry transgenes into cells. Vectors usually have a selectable marker or reporter gene for identifying genetically modified cells (Figure 6–9).

Scientists generally need to make and store many copies of recombined DNA, and this is accomplished with the help of laboratory bacteria. Whenever bacteria grow and divide, they replicate their chromosomal DNA and any plasmid DNA that has an **origin of replication** sequence, passing all this genetic information on to daughter cells. Plasmids used in genetic engineering are placed in bacterial hosts, and the bacteria make copies of the plasmid DNA during

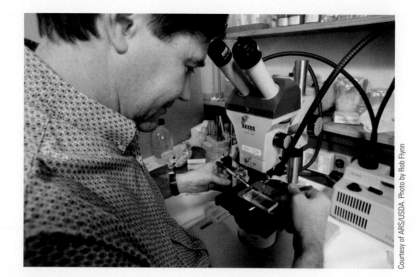

Figure 6-9
This scientist is injecting a vector containing a marker into insect eggs.

Courtesy of ARS/USDA. Photo by Rob Flynn

cell division. This process, called **cloning**, results in many thousands of identical copies of the DNA of interest. In this way, scientists use the bacterial cells as "plasmid DNA factories." When the bacterial cultures reach optimal stages of growth, the cells are collected and broken open, and the plasmid DNA is purified and stored in a freezer for later use in genetic engineering experiments.

Cell Transformation

Once a desirable vector has been created, how do scientists move that DNA into cells? As you may recall from Chapter 4, all cells are enclosed by a cell membrane, and some cells also have a cell wall. In order to deliver recombinant DNA into cells, a passage through the cell membrane and, possibly, a cell wall must be created. There are a couple of artificial ways to open up pores in the cell membrane without killing the cell. Bathing bacterial cells in calcium chloride salts will make them able to receive vector DNA. Once treated, these cells are called competent cells. **Competent cells** are then mixed with the vectors and are exposed to a quick heat shock or electrical pulse (**electroporation**) to create pores in the cell membrane. The plasmid DNA enters the cell through these pores. Plant protoplasts (cell wall removed) and animal cells may undergo similar procedures to take up DNA.

Scientists have also found ways to adapt certain viruses and bacteria as recombinant DNA "delivery trucks." These viruses and bacteria have natural mechanisms for puncturing the cell membranes of target host cells and delivering genetic information. This is the process by which viruses and bacteria infect healthy cells. We can "disarm" these pathogens by taking away their genes for causing infection and substituting genes of interest to be delivered. To find a particular gene responsible for a particular characteristic, a scientist must locate what are called **markers**. Markers are generally found by comparing DNA patterns on fragments of DNA (Figure 6–10). Samples of DNA from an organism having the desired characteristic are compared to DNA fragments from a very similar organism not possessing the desired characteristic. For example, if a scientist wishes to locate the gene responsible for giving a corn plant resistance to a virus, he or she will compare a fragment of DNA from a plant having the resistance to fragments of DNA from a plant not having the resistance. Differences in the samples of DNA fragments are examined and noted. These differences can be used as markers to locate the desired gene. The comparison of these patterns is known as **restriction analysis**. A slightly different form of restriction analysis is also used. This technique is called **restriction fragment length polymorphism**. In this process, a scientist will cut fragments of DNA from tissue to be compared using an enzyme that cuts DNA at a specific site. These enzymes are called restriction enzymes. The sample fragments are then sorted by a process called **electrophoresis**, a method of sorting DNA fragments according to size by subjecting the fragments to a direct electric current that draws the DNA through a gel. The negatively charged DNA is sorted because the smaller pieces of DNA move more rapidly through the gel. After the DNA is sorted by

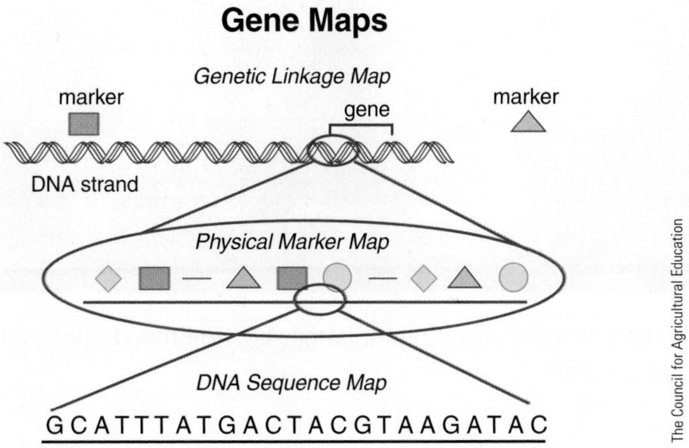

Gene Maps

Genetic Linkage Map

marker   gene   marker

DNA strand

Physical Marker Map

DNA Sequence Map

G C A T T T A T G A C T A C G T A A G A T A C

The Council for Agricultural Education

Figure 6–10
Gene markers are located along strands of DNA.

Scientists Use Old, New Tools to Develop Pest-Resistant Potato

Despite their microscopic size, Columbia root-knot nematodes (CRN) have potential to inflict huge losses—about $40 million annually—by tunneling into potatoes to feed. But this level of loss is not likely to happen, thanks to fumigants growers now use—at a cost of $20 million annually.

In seeking alternatives to using chemical fumigants, ARS and collaborating scientists are field-testing a new russet potato breeding line that naturally resists the pests.

Commercial varieties bred from line PA99N82-4 would be the first with resistance not only to CRN, but also to northern and southern root-knot nematodes, says geneticist Chuck Brown. He is in ARS's Vegetable and Forage Crops Research Unit at Prosser, Washington. "PA99N82-4 also resists the viral disease corky ringspot, which is transmitted by nematodes and causes unsightly blemishes in tubers," he adds. "Corky ringspot is also controlled by soil fumigation."

CRN is problematic in the Pacific Northwest, where two-thirds of America's potatoes are grown, and in Florida. Though fumigating the soil before planting suppresses CRN numbers, the practice is not cheap, with some chemicals costing $300 per acre. It can also harm nontarget organisms, including beneficial soil-dwelling insects.

Genetic resistance, however, confines the fight to the potato's roots and tubers. But putting that resistance to work has not been easy.

Because resistance is absent from United States cultivated potatoes, Brown and colleagues used the wild species collection at

A potato infected with Columbia root-knot nematode (left) and a healthy potato.

ARS's U.S. Potato Genebank in Sturgeon Bay, Wisconsin. Painstaking screening of the material at Prosser showed *Solanum bulbocastanum* to be the most resistant.

The problem is, wild and cultivated potatoes are chromosomally incompatible. So the researchers resorted to "bridging," a technique that fused *S. bulbocastanum* and cultivated potato cells together, forcing the DNA of both to combine. Stimulants were then added to induce cells to become plantlets. Over several years, the researchers used backcrossing to eliminate unwanted traits—like tiny tubers and poor taste—from resistant plants they had created.

Besides conventional plant-breeding techniques, they used biotechnology methods, including DNA markers linked to *S. bulbocastanum*'s gene for resistance, *RMc1(blb)*. Normally, resistance levels are determined by inoculating potted plants with nematodes, waiting 7 weeks, and removing and washing the roots so the pests' eggs can be counted.

"It's a laborious, time-consuming process," says Brown. But with marker technology, leaf tissue can be quickly analyzed for genetic evidence of *RMc1(blb)*. "Being able to determine in 1 day's time which plants are resistant is very helpful," he adds.

Still, the entire process to date has taken 20 years and the close collaboration of many scientists, including ARS postdoctoral researcher Lin-Hai Zhang, at Prosser; Washington State University scientist Hassan Mojtahedi; and John Helgersen, now retired from ARS. PA99N82-4, the top pick of this intensive effort, is in its third year of field trials. Besides tests in Washington, Oregon, and Idaho under the Tri-State Potato Breeding Program, it is also being evaluated in California and Texas. Two more years of testing will follow before the line is released for development into commercial varieties.

—*By Jan Suszkiw, Agricultural Research Service information staff.*

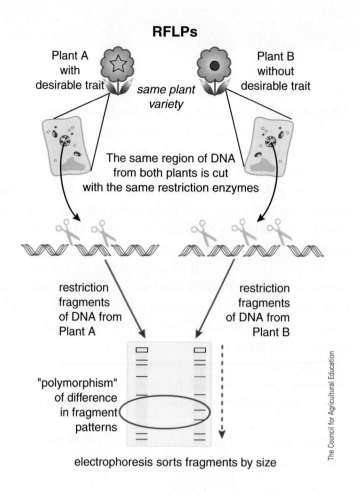

RFLPs

Plant A with desirable trait

Plant B without desirable trait

same plant variety

The same region of DNA from both plants is cut with the same restriction enzymes

restriction fragments of DNA from Plant A

restriction fragments of DNA from Plant B

"polymorphism" of difference in fragment patterns

electrophoresis sorts fragments by size

The Council for Agricultural Education

Figure 6–11

A process called electrophoresis sorts fragments of DNA by size.

size, the double strands of DNA are split into single strands and are then placed on a membrane (Figure 6–11). A radio-active **DNA probe** is then added, and the probe adheres to complementary sequences on certain fragments. The DNA fragments that do not adhere to the probe are simply washed away. The fragments of DNA that are attached to the radio-active probes are exposed to photographic film where an image of the DNA appears (Figure 6–12).

In isolating and cutting DNA for splicing into an organism, it is not enough to merely use the gene responsible for the trait. Gene sequences known as **controller genes** are also necessary. These genes are required to regulate when the gene is activated or inactivated. For example, a gene for milk production must also have a controller sequence, or promoter, that tells the lactation system when to produce milk and when to stop lactating. These promoter sequences may work in concert with hormones produced by the animal.

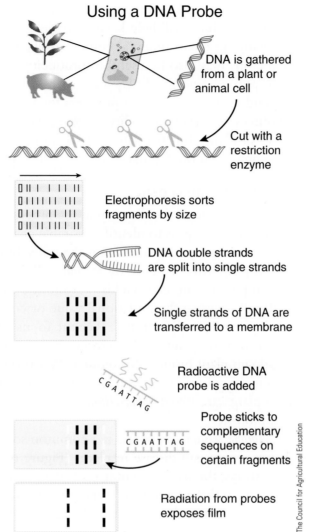

Using a DNA Probe

DNA is gathered from a plant or animal cell

Cut with a restriction enzyme

Electrophoresis sorts fragments by size

DNA double strands are split into single strands

Single strands of DNA are transferred to a membrane

Radioactive DNA probe is added

Probe sticks to complementary sequences on certain fragments

Radiation from probes exposes film

Figure 6–12
Radioactive probes are used to locate DNA sequences.

The Council for Agricultural Education

Genetic Modification in Plants

As will be discussed in detail in Chapter 9, a significant number of staple crops produced by modern agriculture have been genetically modified. The most common genetically modified crops are soybean, maize, cotton, and canola. There have also been a number of specialty crops, such as papaya and squash, the have benefited from biotech innovations. By acreage, the most common traits are insect resistance and herbicide tolerance. There are also crops in development that have been nutritionally enhanced, such as Golden Rice and BioCassava. Crops with traits for stress tolerance to drought, saline soil, and temperature extremes and those that use less fertilizer will help keep crop yields high in the face of changing environmental conditions. Genetically modified crops may also be useful in producing useful proteins, such as medicines, and play a major role in alleviated our dependence on petroleum products for fuel and industrial resources. We have the ability to genetically engineer many types of plants, but regulatory, legal, and economic factors have prevented the full usage of the technology, so far.

Transforming Plant Cells

As outlined above, the first step in the process of genetically engineering crops is to identify desirable characteristics and the gene or genes responsible for the traits. Interesting traits for crop improvement are those that will decrease production inputs or increase yields (agronomic traits) or those that will improve the nutritional value or consumer appeal of the crop (quality traits). Recent focus has also been given to using plant cells as protein bioreactors and breaking down plant biomass for biofuels. Whatever the goal, the two most common ways to introduce recombinant DNA into plants are via *Agrobacterium*-mediated transformation and biolistics.

Agrobacterium tumefaciens is a common soil bacteria that causes crown gall disease in plants (Figure 6–13). In natural settings, the bacteria enter host plants through wounds and sets up the infection by using plasmids to deliver genes into the plant cell genomes. The genes on the plasmid

Figure 6–13

A bacteria that causes tumors on plants is used as a vector to implant DNA.

Courtesy of Dr. Jean L. Williams-Wooward, The University of Georgia encode proteins that cause plant cells to swell, providing a living space for the bacteria, and to produce molecules, which the bacteria use as food. All in all, it is a very good example of pathogen exploitation of its host. Understanding this interaction has proven very useful for plant scientists. Scientists have reengineered plasmids in the *Agrobacterium* and used it to deliver genes of interest into plant genomes. This method does not work well with monocots such as cereal crops; however, it works well with dicots such as soybeans.

Whole plants at the flowering stage or pieces of plant tissue, such as leaves, may be transformed using *Agrobacterium* as the method of transformation. With whole plants, the hope is that reproductive cells within the flowers will be transformed. Seeds of transformed plants are germinated and offspring screened for the inserted genes using selectable markers and reporter genes. With transformation of plant tissues, whole transgenic plants are regenerated through vegetative propagation (covered in detail in Chapter 8). Transformed plant cells are placed on a selective growing medium, perhaps containing antibiotics, that allows only the genetically modified cells to grow. These cells are then regenerated into mature plants in a carefully controlled environment using proper hormones, nutrients, temperature, and humidity. This process is known as **tissue culture**. Genetically modified plants produced via vegetative propagation in tissue culture are eventually transferred to soil and grown for seed.

Biolistics literally blasts recombinant DNA into plant cells using tiny bullets composed of tungsten or gold particles, coated with the DNA of interest. Usually a .22-caliber gun cartridge is used to fire the particles, but compressed air may also be used (Figure 6–14). Although some of the cells are destroyed in the process, some of the cells remain intact with the fragments of DNA embedded. Once inside the cells, the DNA integrates into the plant's genome. Transformed tissues are cultured, and whole plants are regenerated in the same manner as described for the *Agrobacterium* method of transformation. The biggest advantage of this technique is that it can be used with intact tissues, such as meristems of plants, or other tissues that are capable of regenerating. Many of the genetically modified crop plants have been created using this technique.

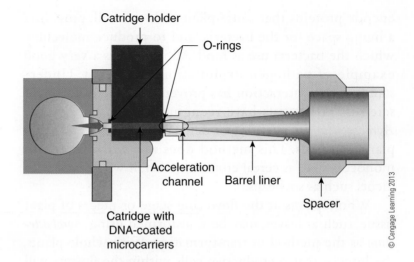

Figure 6–14

A gun using compressed air with a .22-caliber cartridge can blast DNA into plant cells.

© Cengage Learning 2013

Transient Expression of Proteins

Researchers have developed a method of temporarily changing the genetic content of plant cells, allowing the quick expression of certain proteins to act as vaccines or pharmaceuticals. The idea is to harvest fresh plant tissue, such as leaves, and use the *Agrobacterium* method of transformation to infect the tissue, delivering DNA that codes for the cauliflower mosaic virus as well as genes of interest. The plant cells then behave as though infected with a virus and express the viral RNA and proteins of interest without integrating this genetic information into the plant chromosome. The expression of these proteins is transient or temporary, and it ceases when the transformed RNA degrades over time.

The advantage of this method is the ability to produce specific proteins very quickly without having to transform and regulate a field-grown genetically modified crop for each different protein. In the case of vaccines, for example, different vaccines may be needed in large quantity at different times in order to respond to disease outbreaks. It is estimated that it would take about 3 weeks to perform the recombinant DNA experiments and transformation of plant tissue necessary to produce a large quantity of vaccine with this method. Currently, vaccine production in eggs and other types of cells is much slower and cannot yield the high quantities that the transient expression of plant cells can yield.

Courtesy of ARS/USDA

Figure 6–15
Transgenic animals hold great promise for the animal industry.

Genetic Modification in Animals

Improving livestock by creating transgenic animals holds great promise for the agricultural industry (Figure 6–15). Because the process of reproductive cloning and genetic engineering in animals is still in development, techniques are ever evolving and rely on our ability to understand the complex animal genomes. To conduct genetic engineering, new genetic material must be introduced into the animal's genome without upsetting the overall genetic program of the animal. In other words, the inserted genetic material must have a benefit that can be expressed without causing harm or adversely affecting the genes that control all the other of the animal's characteristics. If the foreign DNA creates problems for the animal, the purpose of the process will be defeated. Scientists must take extreme care in the selection of the foreign DNA introduced into an animal's genome.

Foreign genetic material must be spliced into the animal's germ cells in order to transfer the desired characteristic to the next generation. This means that the inserted

DNA must be placed into the sperm cell, egg cell, or embryo in its early stage of development. Unlike plant cells, the new animal must be incubated inside the mother's womb instead of a test tube. This can create problems with embryos containing foreign genetic materials because the mother's reproductive system tends to reject or abort fetuses that are not normal. For a number of other physiological reasons, the success rate of producing transgenic offspring is often less than 10 percent. Producing genetically altered animals may take many tries before success is achieved.

Most of the transgenic animals produced to date have been mice because they reproduce so quickly and can be easily maintained in the laboratory (Figure 6–16). Mice also serve as an important model species for genetic studies and pharmaceutical trials. However, many other genetically modified animals have also been successfully produced, including cattle, pigs, goats, poultry, and fish.

Techniques for Creating Transgenic Animals

As in the gene splicing process in plants, the first steps in animal transformation are the selection, location, and preparation of the desired genes. This is accomplished using techniques very much like those used in locating and isolating plant genes. However, the techniques for transferring the recombinant DNA to the cells are slightly different.

Figure 6–16

Most transgenic animals are mice because they reproduce so quickly and can be easily studied in the lab.

The first method developed to insert genetic material into animal cells involved using a device called a micromanipulator that is mounted on a modified microscope (Figure 6–17). Using this **microinjection** method, fertilized eggs (early "two-pronuclei" zygotes) are removed from an animal and placed under a microscope. The micromanipulator is used to grasp the **zygote** and the genetic material is inserted using an extremely fine injection needle (Figure 6–18). The DNA is placed into one of the zygote's

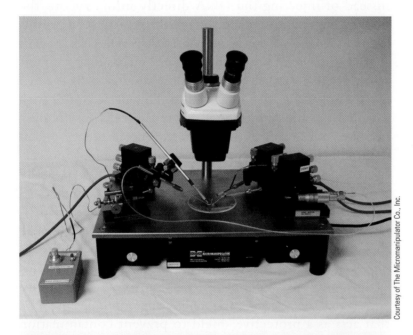

Figure 6–17
Scientists use a device called a micromanipulator to place DNA into animal cells.

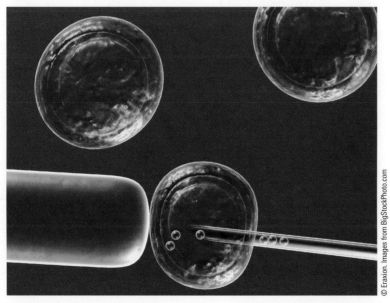

Figure 6–18
The micromanipulator grasps the zygote and inserts genetic material using an extremely fine needle.

two pronuclei. A **pronucleus** is either the nucleus of the sperm or the egg that contains half the chromosomes of the fertilized ovum. Pronuclei eventually fuse as the zygote begins to develop. The zygote is then placed back into its mother's uterus where it will hopefully develop into a full-term transgenic animal. With this method, only about 4 percent of genetically modified zygotes develop into full-term newborn animals.

Another method also uses the micromanipulator, but instead of injecting the DNA directly into a zygote, the genetic material is placed in **embryonic stem cells**. Remember from a previous chapter that stem cells are cells that are capable of producing any of the cells in an animal's body. These are taken from an embryo after it has grown for about 7 days after fertilization. This period of growth is called the **blastula stage**, where the cells are dividing but have not begun to differentiate. One advantage of this method is that the embryonic cells can be altered and cultivated in the lab to produce a large number of genetically modified cells for use in generating transgenic animals.

A third method, called **transgametic technology**, involves the use of a viral vector. The virus is specially treated to remove any harmful effects, and a desirable gene is inserted. The vector is then injected into an unfertilized egg. A difficulty encountered in inserting foreign DNA into eggs is that an egg has a membrane surrounding the DNA that acts as a protective shield to prevent contamination (Figure 6–19). This problem is solved by inserting the foreign DNA into the egg before the membrane surrounding the egg has fully developed. Once the foreign DNA is integrated into the egg, the egg is fertilized in the lab via a process called **in vitro fertilization (IVF)**, and the resulting embryo is transplanted into a host mother. Being able to transplant a healthy embryo into the reproductive tract of a female greatly increases the likelihood of the embryo growing to full term and being born healthy.

Creating Knockout Animals

The purpose of the first transgenic animals was to incorporate desirable foreign genes into the genome of a particular animal in order to give the animal a special useful benefit. However, another huge advantage of transgenic animals was discovered—that of genetic research. By replacing an original

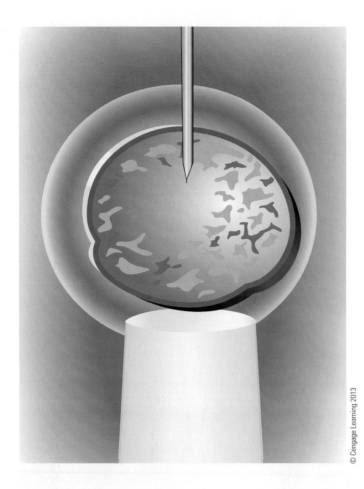

Figure 6–19
A difficulty encountered in inserting foreign DNA into eggs is that an egg has a membrane that acts as a protective shield to prevent contamination.

gene in an animal's genome with an inactive foreign gene, the effects of the displaced gene can be observed. If the original gene is replaced and is no longer functioning, the characteristics of the animal can be compared to those of similar animals with all of their original genetic makeup intact.

The problem has been that foreign DNA is usually inserted at random on the chromosome anyplace where the coding will match. This problem was solved with a process called **homologous recombination** (*homologous* means "arranged in the same place or position"). In other words, a gene for a particular characteristic has a specific place or position on the DNA strand on the chromosome (Figure 6–20). If a new inactive gene is placed in the exact position on the chromosome as the original gene, the new gene does not function, and the original gene is replaced. In other words, the original gene has been effectively "knocked out" of action.

Courtesy of ARS/USDA #K9389-3

Figure 6–20
This scientist is looking for a specific location of a particular gene.

The animal is then referred to as a **knockout animal**. This allows scientists to study what effect the original gene has on similar animals. The process of using homologous recombination holds a tremendous amount of potential for studying the effects and functions of particular specific genes.

Summary

In the view of many scientists, the development and use of GMOs will continue to be a critical part of agriculture, medicine, and industries that make use of living organisms. Since the 1970s, molecular tools and techniques for genetic engineering have been discovered and refined. As we learn more about the regulation of our genes and genomes and about the functions of our cells, tissues, and organs, the production of transgenic plants and animals will become easier and more routine. To spur innovation in this area of research, many scientists feel that we should make sure that core enabling technologies, like those for recombinant DNA work and cell transformation, are accessible to all researchers, both in the United States and in developing countries. With scientists from many backgrounds engaged in the process, genetic engineering may allow us to make the most of natural resources and solve global challenges in health care, agriculture, and the environment.

CHAPTER REVIEW

Student Learning Activities ....

1. Imagine how you might genetically modify a plant to improve the plant. What new characteristics would the plant have? From where would the new genes come? What problem would the transgenic plant solve? What dangers might be encountered with releasing the new plant for production? Share your thoughts with the class.

2. Go to the library or Internet and locate information on a transgenic animal. What genes were placed in the animal's genome? What purpose does the animal serve? List the steps taken in producing the animal. Report to the class.

3. Locate an article where the author is critical of genetically modified organisms. What points does the author make? Do you feel these criticisms are justified? Why or why not? Share your article with the class.

4. Interview 10 people and ask them their views on GMOs (genetically modified organisms). Report your findings to the class.

Fill in the Blanks ....

1. In animals, the advent of methods such as _____ _____ and _____ _____, coupled with computerized _____ _____, have allowed great strides in animal improvement.

2. Organisms produced by _____ DNA technologies to _____ genetic materials between different types of _____ are called genetically modified organisms or _____ _____.

3. Genes for many different types of useful _____, such as therapeutic _____, _____, and industrial _____, have been genetically engineered into bacteria and fungi used in _____.

4. _____ have circular-shaped pieces of DNA called _____ that float freely in the cell's _____.

5. _____ control _____ of genes by interacting with the _____ machinery of the cell.

6. During _____ screening, to increase the chance of finding an _____ trait, a mutation population is often generated using _____ or _____ _____.

7. A group of _____, called the _____ _____, recognize specific DNA sequences and _____ the DNA into pieces.

8. In a process called _____, plasmids used in genetic engineering are placed in bacterial _____ and the bacteria make _____ of the plasmid DNA during cell _____.

9. _____ are generally found by comparing DNA _____ on fragments of DNA from an organism having the _____ characteristic to DNA fragments from a very _____ organism not possessing the desired characteristic.

10. _____ literally blasts _____ DNA into plant cells using tiny bullets composed of _____ or _____ particles, coated with the DNA of interest.

True or False

1. Currently, the use of genetically modified organisms is widespread through the agriculture and medical industries.

2. Genes of interest that are moved into an organism using recombinant DNA technology are called recombigenes.

3. The first organisms to undergo gene splicing and be useful transgenic organisms were bacteria.

4. Cultures of bacteria, fungi, insect, plant, and animal cells can all be used for biomanufacturing proteins.

5. It is harder to genetically engineer prokaryotes than it is to genetically engineer eukaryotes.

6. In reverse genetics, scientists observe the physical characteristics to find organisms with interesting or useful characteristics.

7. Competent cells are cells that have been treated so that they will accept a vector.

8. During electrophoresis, DNA fragments are sorted by size.

9. We are using transgenic technology to its fullest extent in the agriculture industry.

10. In order for foreign genetic material to be expressed in the next generation of animals, it must be inserted into the germ cells (egg or sperm) or in the embryo during early developmental stages.

Discussion

1. What is a genetically modified organism?

2. Compare and contrast selective breeding and genetic engineering.

3. Explain the role of enzymes in genetic engineering.

4. Why was *Escherichia coli* a good choice for biomanufacturing insulin?

5. Explain the process of biomanufacturing insulin.

6. List the questions that must be considered in order to genetically engineer organisms.

7. Explain how plasmids can function as vectors.

8. Describe the process of Restriction Fragment Length Polymorphism (RFLP).

9. What is tissue culture?

10. List and describe two techniques for creating transgenic animals.

CHAPTER 7

Animal Cloning

KEY TERMS

clones
surrogate mothers
genetic improvement
genotype
phenotype
endangered species
genetic code
oocyte
zygote
differentiation
nuclear transfer
enucleated oocyte
quiescent
genetically altered clones

OBJECTIVES

When you have finished studying this chapter, you should be able to:

- Discuss the benefits of cloning animals.
- Describe animals that have been successfully cloned.
- Explain the difference in cloning derived from embryos and cloning derived from differentiated cells.
- Explain the process of nuclear transfer.
- List the benefits of genetically altered clones.
- Discuss how the cloning of cattle has been made more efficient.
- Discuss why there can be differences in animal clones.

Introduction

Throughout literary history, a common theme of science fiction has been the cloning of humans and animals. The plot usually follows the story line that cloning experiments go terribly wrong, and the resulting creatures develop into monsters that create murder, mayhem, and havoc until they are ultimately destroyed (Figure 7–1).

In the past, these stories were considered to be just the rich imagination of science fiction authors, and few people really thought cloned creatures would ever be a reality. However, this thought was abandoned in 1996 when a sheep named Dolly was cloned from a nonembryonic cell of an adult sheep. Suddenly, the topic of cloning dominated the media, and concerns were immediately raised about the ethics and dangers of creating organisms through such an "unnatural" means.

Figure 7–1
In science fiction, cloning always results in the creation of monsters.

Despite the controversy surrounding animal cloning, research and development of this technology has continued at an ever-increasing pace. Large mammals, such as cattle, sheep, and horses, have already been cloned and have received a lot of media attention. Several large companies are investing a huge amount of time and resources into the development of animal cloning. Cloning has become much more common in recent years, with most major species being successfully cloned. The procedure is still quite expensive and inefficient, though improvements are being made as research provides a better understanding of the process.

Why Do We Want to Create Clones?

Although there has been much publicity over cloning, **clones** have always been in our midst. In fact, cloning actually happens in nature. All naturally born identical twins are actually clones as they have the same genetic makeup—and this includes humans (Figure 7–2). If cloning happens in nature, why all the interest in artificially creating clones? What is so special about producing animals with the same genetic makeup? There are several very sound reasons to invest so much money and effort in developing the ability to produce cloned animals.

© Corbis

Figure 7–2
All natural-born identical twins are actually clones.

Genetic Superiority

One of the main reasons for cloning is to take advantage of genetic superiority. Even with the technology of artificial insemination and embryo transfer, the process of genetic improvement in animals can be slow. Through the use of cloning, each cell in the body of a superior animal could theoretically produce a new animal with the same gene pool. If these cells can be harvested, grown into embryos in the lab, and transplanted into **surrogate mothers**, **genetic improvement** can take place much more rapidly.

From studying the chapter on genetically modified organisms, you may recall that the process of inserting DNA into animal cells is a very complicated, time-consuming process (Figure 7–3). If the cells from a genetically engineered animal could be cloned, the process could be made a lot more efficient. One of the problems with selective breeding is that unwanted traits can sometimes be passed along with the desirable traits in superior animals. By carefully selecting the genes that carry only desirable traits, this problem could be eliminated. Theoretically, through the process of genetic engineering, a "superclone" could be developed that would be an ideal agricultural animal. This animal could then be reproduced through multiple clones.

One issue that critics have raised is that cloning is merely reducing the genetic variation in any given population. Their contention is that if all animals in a herd are

Figure 7–3
Inserting DNA into animal cells is a very complicated, time-consuming process.

genetically identical, there will not be enough variation in the population should some problem, such as a new disease, arise.

Animal and Product Uniformity

Genetic diversity has many advantages; however, a drawback is that genetically diverse animals raised together are given care and nourishment aimed at the *average* animal in the flock or herd. This means that there will be animals that need less nourishment or medication and that there will be animals that need more. If all the animals were genetically identical, the environmental care, medications, feed, and other management techniques could be tailor-made for the entire group of animals in the flock or herd.

Animal products from cloned animals could also be more uniform. If an entire flock of cloned chickens were raised in the same environment with the same management practices, the resulting chicken carcasses should be uniform. All drumsticks would be very close to the same size as well as all the other parts of the chicken (Figure 7–4). The same would be true for any of the agricultural animals. If a particular market wanted small T-bone steaks, then cloned cattle that matured early and produced relatively small T-bone steaks could be raised in the same pen, managed the same, and sold at the same time through the same market. Both retailers and consumers could order steaks of a uniform size of their choice. This could make the process of raising meat more efficient and ultimately benefit the consumer by lowering the prices and increasing the quality of meat products.

In recent years, beef animals have been cloned from U.S. Department of Agriculture (USDA)—graded carcasses (Figure 7–5). Researchers toured slaughter facilities throughout the country collecting tissue samples from the highest-quality beef carcasses they could find. In the laboratory, these animals were successfully cloned and have been entered into breeding programs. The theory is that the genetic potential that resulted in a high-quality carcass has been captured and is again available in a new animal. There are several obvious advantages to this method. The first is that a dead steer is reborn as a living bull, with the

© Elena the Wise. Image from BigStockPhoto.com.

Figure 7–4
Drumsticks from cloned chickens could be more uniform in size.

Figure 7–5
Tissue from beef carcasses has been used to clone new animals that were successfully entered into breeding programs.

© Svitlana10. Image from BigStockPhoto.com.

potential to pass along his genes to his offspring. The second is that the carcass merit is definitely known, something that is merely estimated when a living animal is evaluated. These carcasses have been officially graded by the USDA. Although the genetic potential (**genotype**) is passed along through the clone to its offspring, the animal is not guaranteed to be identical, as environmental factors play a large role in the final characteristics (**phenotype**) of the animal.

Endangered Species

Many scientists think that cloning may be a solution to the complete extinction of animal species on the verge of dying out. Over the world, literally hundreds of different animal species are in danger of becoming extinct. Even with the best of our conservation efforts, whole species of animals continue to disappear. If new animals could be reproduced from the tissue of the few remaining animals, many **endangered species** might be saved. In fact, scientists have dreamed of bringing back long-extinct species through the cloning of preserved DNA. For example, woolly mammoths and mastodons have been extinct for thousands of years, yet frozen remains of these animals are periodically discovered in the arctic regions (Figure 7–6).

© dbvirago. Image from BigStockPhoto.com.

Figure 7-6
For many years, scientists have dreamed of bringing back the wooly mammoth by using frozen DNA.

The thought is that if enough intact DNA is preserved in the remains, an embryo of the animal might be cloned by using elephants as surrogate mothers. So far, there has been no success with this goal, primarily because no DNA has been found that is complete and sound enough to be used for cloning.

Some success in the cloning of endangered species has already been achieved. In 2000, an endangered species was successfully cloned. The Asian gaur is an oxlike animal native to India and Burma, which has been endangered for several years. The adults can reach a mature weight of over a ton and have been a favorite game animal for many generations of hunters. Overhunting and the destruction of their habitat have caused the populations of these animals to decrease drastically to the point of extinction.

Using cow eggs with the nucleus removed, DNA from the skin cells of a dead gaur was implanted into the eggs, and the resulting embryos were placed in the reproductive tract of cows (Figure 7–7). All together, 42 embryos were implanted into 32 cows; however, only one live birth of a real, cloned gaur resulted. Although the cloned calf, named Noah, died only 2 months after birth, scientists proved that the technology was possible. Similar research is being performed using domestic sheep eggs to clone endangered species of wild sheep. In the future, genetic scientists and wildlife biologists will attempt to clone many other endangered species.

Figure 7–7
The gaur, an endangered species, has been successfully cloned.

Figure 7–8
Mules are naturally sterile. They have been successfully cloned.

Cloning Sterile Animals

Another potential advantage of cloning is the ability to reproduce animals that are otherwise unable to reproduce themselves. In 2003, Dr. Kenneth White of Utah State University and Dr. Gordon Woods of the University of Idaho were the first to successfully clone a mule, which would otherwise be unable to reproduce itself (Figure 7–8). A mule is the sterile offspring of a mare and a jackass. In this case, the mule was a valuable racing animal, and it was reproduced through cloning.

The same principle has been applied to the cloning of steers, which are born an intact bulls and could enter the breeding population as clones. Cloning could also be of benefit to horse breeders who have particularly valuable geldings that are no longer able to reproduce but could pass on their genetic potential through cloning (there have been nine geldings that won the Kentucky Derby but have not been able to pass on their own valuable genetic potential).

Research

Perhaps the greatest advantage of cloned animals is their use as research animals. Recall from the chapter on conducting research that one of the biggest problems facing researchers is that of controlling all the differences between animals within a group or between groups. A group of animals selected for research will have different genetics even though they may be closely related. Often, this may interfere with the findings of a research study. For example, differences may show up that are more related to genetics than to the treatment given to the experimental group (Figure 7–9). This genetic difference is usually controlled by placing a large number of animals in both the experimental and the control group. Having to use large numbers of animals and having to repeat experiments many times greatly increases the cost and amount of time involved in conducting animal research. However, if genetically identical animals could be produced through cloning,

© C. Sherburne/Photolink/Photodisc/Getty Images

Figure 7–9
One problem encountered in conducting animal research is the genetic differences in animals.

not only could the number of animals needed for the study be decreased, but the results could also be more meaningful. The differences due to genetic variation among animals in an experimental group could be eliminated.

The Animal Cloning Process

For many years, scientists have known that all the cells in an animal's body contain the **genetic code** for the entire animal. This code is created when half the code is passed from the father and half is passed from the mother when the sperm and egg are united during fertilization. When the egg, called an **oocyte**, is fertilized, it becomes a **zygote** with the complete genetic code intact (Figure 7–10).

Remember from Chapter 4 that the zygote begins to divide into identical cells and that the process of cell division continues for 10–12 days until a ball-shaped mass of cells called a morula is formed. At this point, cells begin to change and develop into different types of cells that will divide and grow to form bones, muscle, skin, and so on. The cells from which the **differentiation** begins are called stem cells. Once the cells begin to differentiate, the genetic coding is locked into place, and bone cells produce nothing but bones, muscle cells produce muscles, and so on, although the entire genetic code is in each cell. The genes that encode traits necessary for a bone cell or a muscle cell

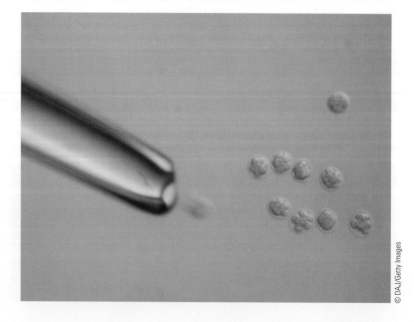

Figure 7–10
When eggs are fertilized, they become a zygote with all the genetic code intact. This is a microscopic image of zygotes.

are turned on, and other genes are turned off. For many years, scientists thought that once the cells differentiated, the process could not be reversed into producing undifferentiated cells.

With this mind-set, the first clones were developed by dividing a zygote to produce two or more animals. They stimulated each half to continue the division process, and clones were born in much the same process as ordinary identical twins, where the zygote divided naturally. It was long thought that this was the only means of creating clones because the process of differentiation could not be reversed.

Frog Clones

Animal clones derived from differentiated cells first became a reality back in 1962 when a scientist named John Gurdon of Oxford University developed a procedure called **nuclear transfer**. Gurdon was able to take the DNA from a cell in the intestine of an adult frog and use this genetic material to clone a frog. He began the process by removing the nuclei from a batch of frog eggs. Remember that the nucleus of a cell is where the DNA is located, so when the nucleus is removed, the DNA is also removed. This results in what is known as an **enucleated oocyte**. From other frogs, Dr. Gurdon took cells from the intestines, removed the nuclei of the cells, and placed them into the enucleated eggs (Figure 7–11).

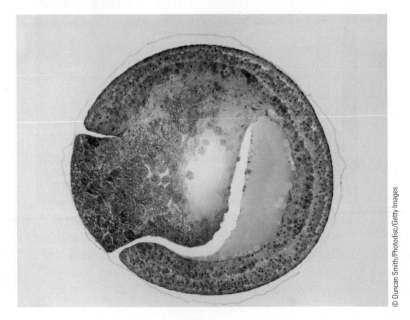

© Duncan Smith/Photodisc/Getty Images

Figure 7–11
A cloned frog was created by placing the nuclei from a frog's intestine into enucleated eggs.

Breeding and Genetic Change in the Holstein Genome

The average American dairy cow produces more than 20,000 pounds of milk every year. Most of these cattle are Holsteins, a breed whose naturally high milk yield has been enhanced by decades of selective breeding.

The U.S. dairy industry intensified selective-breeding efforts in the 1960s. Since then, the average milk yield in Holsteins has doubled, but there has also been a substantial reduction in fertility. Recent research by scientists at the Agricultural Research Service's Animal and Natural Resources Institute (ANRI) in Beltsville, Maryland, and the University of Minnesota (UM) suggests there may be a genetic connection between increased milk yields and reduced fertility. Since the 1960s, UM scientists have maintained a herd of cattle that were never exposed to the selective-breeding practices used by the U.S. dairy industry. As a result, their genomes represent a snapshot of a time before dairy cattle selection efforts intensified.

Scientists in ANRI's Bovine Functional Genomics Laboratory and Animal Improvement Programs Laboratory teamed up with UM colleagues under the leadership of UM geneticist Yang Da to compare the genomes of modern Holsteins with those of the UM cattle.

Although scientists have significantly improved their understanding of animal genetics over the past 250 years, questions about how long-term breeding influences genome structure are still largely unanswered. Fortunately, improvements in storage, preservation, and shipping technology make it easier to maintain and transport genetic material today than at any other time in history—and that makes comparative studies involving historic DNA more practical than ever.

The investigation involved about 50,000 genetic markers known as "single nucleotide polymorphisms," or SNPs (pronounced "snips"), drawn from about 2,000 cattle. ANRI geneticist John Cole coordinated collection of Holstein

Geneticist Tad Sonstegard analyzes BovineSNP50 BeadChips for genotypic data that decodes each animal's genome at more than 50,000 locations. This type of data is used in cattle research ranging from genome selection to mapping of congenital defects.

DNA samples from the ARS National Center for Genetic Resources Preservation (NCGRP), the U.S. Holstein Association, and five U.S. universities, including UM.

Genome Analysis

Under the leadership of ANRI geneticist Tad Sonstegard, the scientists extracted DNA and genotyped the samples with an Illumina Bovine SNP50 BeadChip, a genetic analysis tool developed by Sonstegard and ANRI geneticist Curt Van Tassell in collaboration with industry, university, and other ARS partners.

The BeadChip is a glass slide capable of generating genetic information that can assist scientists in comparative genetic studies. Each BeadChip is designed to characterize 12 different samples, and each sample is analyzed at more than 50,000 SNP marker locations, referred to as "loci." The scientists decode all the SNP marker information to determine the specific allele combinations at each locus for every sample. In this case, an "allele" represents one of the two detectable forms of a gene. DNA information gleaned from the BeadChip allows scientists to determine which allele was inherited from the mother and which was inherited from the father. An animal can receive the same allele type from both parents or a different allele from each parent. Differences in alleles can result in different characteristics.

"A decade ago, this kind of genotyping work was much more expensive and time consuming," Van Tassell says. "The BeadChip makes it possible to get the information faster and more economically."

The scientists analyzed the allele data to determine which regions of the genome were most likely to have been affected by the selection practices used by the dairy industry to improve milk production over the past 40 years.

The researchers are now focused on six genomic regions that had been identified as being associated with milk yield. The genomic changes made evident by the analysis were "extensive," Sonstegard says. As much as 30 percent of the Holstein genome may have been influenced by standard breeding practices.

"Small changes in allele frequency do occur randomly," Van Tassell says. "But significant changes like these are rare and are almost certainly the result of selection."

The scientists observed that many of the genes and chromosome regions associated with milk yield were also related to lowered fertility rate, supporting their hypothesis of a genetic link between the two traits. As high yield and high fertility are both desirable traits, it is important for producers to recognize that there may be a tradeoff between the two.

"Ultimately, this work may help us understand the limitations of biology as we work to breed a high-performance dairy cow," Cole says.

—By Laura McGinnis, formerly with ARS.

Through this process of nuclear transfer, he created many cells with new nuclei. Although most of the cells died, some of the nuclear transfer cells began to behave in much the same way as a fertilized egg. These cells began to divide, and after a while a morula formed. From this structure, the cells began to differentiate and develop into tadpoles.

For the first time, a scientist had demonstrated that cell differentiation could be reversed. However, a large problem arose from the experiments. Even though the tadpoles appeared normal, they never developed into frogs. Other scientists duplicated Gurdon's research, but no one could get the cloned tadpoles to develop into grown frogs. The reason remains a mystery.

Gurdon's process was tried many times with other animals. Scientists were particularly interested in cloning mammals such as pigs and cattle because of the usefulness and economic value of these animals. However, no one was able to get nuclear transferred mammal cells to grow and divide beyond a few cells. The scientific thought at the time (1970s and 1980s) was that for some reason the process would not work in mammals as it had in tadpoles. However, some scientists refused to give up and continued to research methods of cloning mammals. They were successful in the late 1980s when sheep, cattle, and rabbits were successfully cloned (Figure 7–12). However, these clones were accomplished by dividing embryos and were not the result of the nuclear transfer of DNA from an adult cell.

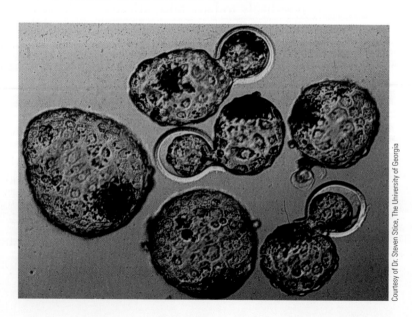

Figure 7–12

These are nuclear transfer, cloned embryos.

Courtesy of Dr. Steven Stice, The University of Georgia

Cloning Mammals

A theory was developed in the 1980s when scientists began to study how cells divide and differentiate. For many years, they had known that some cells divide more rapidly than others. The theory is that cells such as those that makeup skin, hair, or inner organs go through their division cycle more rapidly than other body cells. This is because repairs have to be made as cells are injured or wear out on such exposed areas as skin. Cells go through periods of inactivity and are said to be **quiescent**. During these periods, cells do not divide because there is no need for rapid cell division. Then, when the cells are needed, some mechanism, still not very well understood, triggers the cells into action.

In the 1990s, scientists at Roslin Institute in Scotland began to experiment with quiescent cells as a way of cloning mammals. They began by taking rapidly dividing cells from the mammary gland of a pregnant white-faced sheep and culturing the cells in the laboratory. As a treatment, they deprived the cells of nutrients to stop them from growing. Using previously developed technology, they removed the nucleus of an oocyte from a black-faced ewe, and the DNA from the white-faced sheep's mammary gland was placed in the cell. A small current of electricity caused the foreign DNA to enter and fuse with the cytoplasm in the enucleated oocyte (Figure 7–13). Nutrients were then added to cause

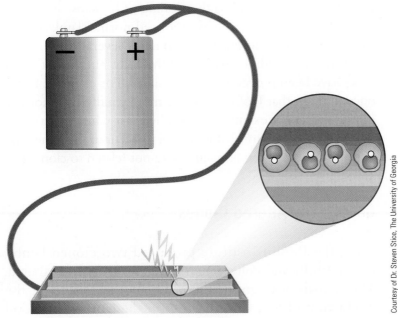

Courtesy of Dr. Steven Stice, The University of Georgia

Figure 7–13
Using electricity, foreign DNA may be caused to enter and fuse with the cytoplasm of an enucleated oocyte.

the new cells to begin to divide. The dividing embryo was then placed in the reproductive tract of a black-faced sheep.

Of course, this was not a one-time process that achieved immediate results. The scientists created 277 new embryos from the process, and of these they recovered only 29 that were good enough to transplant into ewes. Thirteen ewes were given either one or two of the embryos, and of these 13 only 1 grew to full term and produced a normal healthy lamb. As the lamb had a white face and was born to a black-faced surrogate mother, the scientists could be almost sure that the lamb was a product of their cloning process. All doubt was removed when DNA fingerprinting proved that the DNA from the cloned sheep, named Dolly, matched the cells from the tissue taken from the mammary gland.

Dolly created quite a phenomenon in all the media. Television and radio newscasts all over the world carried stories about the birth of Dolly. The Roslin Institute estimated that in the first week after the story broke, it received over 2,000 phone calls about the clone. In addition, during that same week, they talked at length to almost 100 reporters, and Dolly was filmed by 16 film crews and photographed by over 50 photographers associated with the media.

Many of the news stories raised ethical concerns about cloning animals. Questions were asked about whether Dolly was a "normal" animal and the type of creature she might develop into. Concerns were also voiced about the likelihood that the research might lead to the cloning of humans. Since that time, Dolly has matured and produced normal lambs of her own, and fears about her have subsided. Several stories were published about abnormalities with Dolly, but almost all of these have been proven to be unfounded. For example, it was once thought that Dolly was aging more rapidly than normal. However, a closer examination revealed that while the arthritis in her joints was unusual, it is known to occur in sheep her age. Speculation is that the problems were not related to cloning.

Genetically Altered Clones

Later, the Roslin Institute produced two cloned lambs, named Molly and Polly, which had been genetically altered. The animals were given a human gene responsible for the production of a protein that aids in the clotting of blood after an injury. The protein, Factor IX, could be very useful as a pharmaceutical in treating human patients who suffer from hemophilia. The idea is that if genetically altered animals can be mass-produced, the substance can be commercially produced at a much lower cost than conventional methods.

The first **genetically altered clones** were produced in 1998. Two calves, named George and Charlie, were developed by Dr. Steven Stice and Dr. James Robl. These researchers placed two genes—a genetic marker and a gene that makes cells resistant to antibiotics—into Holstein cattle DNA that was used to clone the two calves (Figure 7–14). Although the main benefit gained from this procedure was research, the process proved that genetically altered calves could be produced. Later, Dr. Stice led efforts to clone calves, possessing a gene to enable cows to produce milk with the human serum albumin. Every year, around 440 tons of human serum albumin are used in hospitals to treat patients. If this process can be made commercially feasible, a tremendous boon to the medical establishment will

Courtesy of Dr. Steven Stice, The University of Georgia

Figure 7–14
George and Charlie were the first genetically altered clone calves.

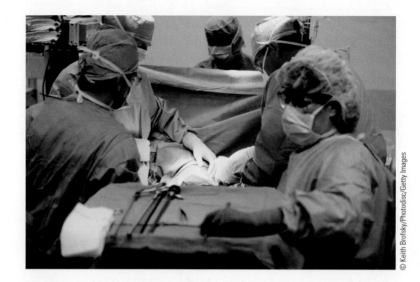

Figure 7–15
Every year, hospitals use hundreds of tons of human serum albumin.

be achieved (Figure 7–15). In addition, it will open the door for many more beneficial products that can be produced in this way.

Another breakthrough occurred in 2002 at the University of Georgia when Dr. Stice succeeded in cloning a calf from cells taken from a cow that had been dead for 48 hours. The cells were taken from a kidney of the slaughtered animal because the kidneys usually remain with the carcass until the carcass is divided into retail cuts. As mentioned previously, this process has been successfully used to clone animals with superior carcass characteristics and allow them enter into the breeding population.

Perfecting the Process

Several times, research has shown that mammals can be successfully cloned. The challenge now is to make the process easier and more efficient. Remember that when Dolly was cloned, 277 cloned embryos resulted in only 1 lamb, and of the 13 ewes that received 29 healthy embryos, only Dolly was carried to full term. Obviously, this success rate is far too inefficient to be of any practical use. If cloning is ever to achieve the potential outlined by futurists, the process will have to be greatly refined and perfected.

In 2001, a team of scientists led by Dr. Steven Stice at the University of Georgia reached a milestone in this effort when eight calves cloned from the same adult cow were born (Figure 7–16). All the calves were born from different surrogate mothers over a period of about 2 months. The significance of this achievement is that in the past, the best viability rate of cloned cattle embryos was about 1 in 20. Dr. Stices's team reduced that rate to one in seven.

The team used eggs harvested from the cattle ovaries obtained from the slaughter plant. Using the eggs as host cells, the nuclei were removed in much the same way as described in previous research (Figure 7–17). The process differed in that a chemical inhibitor was applied to the cells from the donor cow. The chemical makes the cell more uniform in preparation for cloning. In other words, the DNA material used for cloning was made more uniform through the use of the chemical inhibitor.

Differences in Clones

An interesting phenomenon about cloning is that there can be observable differences in clones. Take a close look at the calves in Figure 7–18. Can you detect any differences in the

Dr. Steven Stice, The University of Georgia

Figure 7–16
All eight of these calves were cloned from the same adult cow.

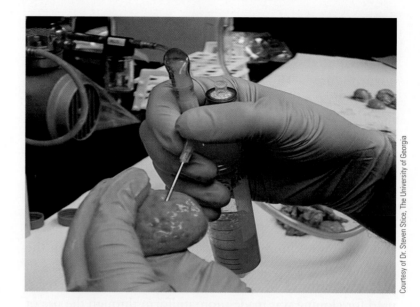

Courtesy of Dr. Steven Stice, The University of Georgia

Figure 7–17
Eggs harvested from cattle ovaries.

Courtesy of Dr. Steven Stice, The University of Georgia

Figure 7–18
Color patterns can be different in cloned calves.

calves? Obviously, the size difference is due to the differences in the ages of the calves. But what about the different color patterns on the animals? As they all have an identical genetic makeup, how can they be so different in the patterns? Remember from Chapter 3 that the genotype of an animal is the actual genetic makeup and that the phenotype is how the genes are expressed, or how the animal actually looks. Environmental factors usually have a larger impact on an animal's phenotype than on its genotype. A study of genetics and heritability indicates that the most highly heritable traits still rely on environment for more than half the measured phenotypic outcome. Remember that all the calves were born to different surrogate mothers. Differences in nutrition, condition of the placenta, or the degree of heat absorbed by the fetus can all affect the color patterns of the individual calves. Also notice that the color pattern on the heads of the calves appears to all be the same. Head markings on calves do not seem to migrate during gestation as do markings on other regions of the body. With all the knowledge we have accumulated about cloning animals, there is a tremendous amount yet to learn.

Summary

Until recently, the cloning of animals has been a fantasy of science fiction writers. In recent years, several breakthroughs in our understanding of DNA and the gene transfer process have allowed scientists to make this concept a reality. Several animals, including rabbits, cattle, and sheep, have been successfully cloned using DNA from adult animals. Many benefits can be gained by cloning animals if the process can be made commercially feasible. Although the technology to make this happen is still just around the corner, most geneticists believe that the techniques can be developed to the point where the cloning of agricultural animals is a common occurrence. Given concerns raised over cloning, it remains to be seen whether the public will accept cloning as part of our society. A discussion of these concerns will be addressed in another chapter.

CHAPTER REVIEW

Student Learning Activities

1. Search the Internet and locate information on an endangered species that is threatened with extinction. Decide whether you believe that cloning technology would benefit this species. What would be the difficulties encountered with this animal? What closely related animals could be used as egg donors and surrogate mothers? Share your findings and conclusions with the class.

2. Conduct a survey of teachers and students in your school to determine attitudes toward animal cloning. Ask if they can give examples of animals that have been successfully cloned. Find out whether they think that scientists should develop and perfect the cloning process. List their comments and share with the class.

3. Choose a species of agricultural animal and write a report on the benefits of cloning that species. If the animal could be genetically altered before it is cloned, what beneficial traits should be added? Share the results with the class.

4. As a class project, prepare a display illustrating the process of animal cloning. Prepare sketches and illustrations on posters and arrange them so that people can understand the process. Place the display in a prominent place in your school. Have someone at the display to answer questions.

Fill in the Blanks

1. The first cloned animal, a _____ named _____, was cloned from _____ cells of an adult sheep in _____.

2. _____ clones derived from _____ cells first became a reality back in _____ when a scientist named John Gurdon of Oxford University developed a procedure called _____ _____.

3. Cells go through periods of _____ and are said to be _____ when there is no need for rapid cell _____.

4. When Dolly was cloned, _____ cloned embryos resulted in only _____ lamb, and of the _____ ewes that received _____ healthy embryos, only Dolly was carried to _____ _____.

5. An interesting _____ about cloning is that there can be observable _____ in _____.

6. If cells can be harvested, grown into _____ in the lab, and _____ into _____ _____, _____ _____ can take place much more rapidly.

7. The process of _____ DNA into _____ cells is a very _____, _____-_____ process.

8. Critics of cloning contend that if all animals in a _____ are genetically _____ there will not be enough _____ in the population should some problem, such as a new _____, arise.

9. While the genetic _____ is passed along through the clone to its _____, the animal is not guaranteed to be _____, as _____ factors play a large role in the final _____ of the animal.

10. A group of animals selected for _____ will have _____ genetics even though they may be closely _____. Often this may interfere with the _____ of a research study.

True or False ....

1. Clones can never occur naturally.

2. Once cells begin to differentiate, their genetic coding is locked into place, and they become a certain cell type such as muscle or bone.

3. Nuclear transfer was first performed using frogs.

4. Nuclear transfer is commonly used to clone adult organisms.

5. The main challenge for scientists at this point is to make the cloning process easier and more efficient.

6. Cloning can produce completely normal animals.

7. Scientists now think they have a complete understanding of the process of cloning.

8. Most scientists agree that cloning animals will never be financially feasible.

9. One of the problems with selective breeding is that unfavorable characteristics can be passed on to offspring along with the favorable characteristics.

10. One disadvantage to cloning is that the animal products produced from clones are uniform.

Discussion ....

1. What is the main disadvantage to animal cloning?

2. List at least two advantages to animal cloning.

3. How can cloning increase genetic superiority in an animal herd?

4. What is a "superclone?"

5. Why is it beneficial to make all the animals in a herd uniform?

6. How might cloning reduce the number of endangered and extinct animals?

7. Distinguish between a stem cell and a differentiated cell.

8. Explain how nuclear transfer works.

9. Why was the cloning of Dolly produce so much controversy?

10. How can clones of the same organism have different physical characteristics?

CHAPTER 8

Plant Cloning

OBJECTIVES

When you have finished studying this chapter, you should be able to:

*   Explain how plants naturally propagate by cloning.
*   List the advantages of plant cloning.
*   Describe the different plant parts associated with vegetative propagation or cloning.
*   Describe the various methods of cloning plants.
*   Explain the role of plant hormones in plant reproduction.
*   Explain why cloning is important in the production of genetically altered plants.

Introduction

Plant cloning is one of the oldest examples of biotechnology. Humans have been cloning plants for hundreds of years as they propagated plants for food, for useful fibers, or for enhancing the environment. In fact, many plants reproduce by means of cloning through a process known as **asexual reproduction** or **vegetative propagation**. Some plants, such as strawberries and many types of grasses, send out runners that grow from the parent plant and develop roots and shoots, forming a new plant (Figure 8–1). The new plant is a clone because it is identical to the parent plant. It was probably through observations of the natural processes of vegetative propagation that humans first began to develop methods of clonal propagation of plants for their own use.

Figure 8–1

Some plants, such as strawberries, send out runners to create new plants. This is an example of plant cloning.

Advantages of Plant Cloning

The cloning of plants is a common process and has many advantages to producers. As the cloned plant is genetically identical to the parent plant, a superior plant can be reproduced without losing the traits that make it superior. For example, if a new variety of apple is developed that tastes delicious and is a high producer, the same type of taste and productivity might be difficult to produce in the next generation if the new plants are produced from seed (Figure 8–2). The reason is that, through seed production, half the genes must come from another plant, and the combination of genes from each

Courtesy of ARS/USDA #K7252-65

Figure 8–2
If a certain variety of apples tastes particularly good, the quality might not be as good in the next generation if it comes from seed.

parent may not yield the same desired characteristics. This may be seen in many new varieties that were produced by crossing plants of different varieties. If, however, the apple can be vegetatively propagated through cloning, it will have the same genetic makeup as the parent and will have the same desirable traits.

Another benefit of plant cloning is that new plants are produced more efficiently through vegetative propagation than those produced from seed through sexual reproduction. To produce seeds, plants must complete an entire cycle of flower production, pollination, seed maturation, and germination. In most cases, this cycle takes an entire year (Figure 8–3). If the new plant is grown from a part of the parent plant, it can be produced without going through these cycles. This means that the grower can produce higher-quality plants in a shorter period of time.

An important commercial development made possible by plant cloning is the production of plants that yield seedless fruit. In nature, the fruits of a plant usually contain seeds that produce the next generation of new plants. However, the seeds are a nuisance when the fruit is eaten. For example, consumers greatly prefer grapes that have no seeds. Through a process of selective breeding, scientists developed several varieties of seedless grapes. The obvious problem encountered was how to produce new generations of plants if the parent plants produce no seed. Through the use of cloning by vegetative means, new grape vines can be produced that will generate grapes without seeds (Figure 8–4).

Figure 8–3
Plants grown from seed usually take a year to produce the next generation.

Courtesy of ARS/USDA #K5632-3

Figure 8–4
Through the use of cloning,
seedless grapes can be produced.

Some plant species, especially fruiting trees, lack uniformity and can vary greatly in quality. Plants having the desired characteristics can be cloned to produce new plants of a high uniform quality. A lot of research is being conducted on the cloning of trees to be used for lumber. Tall, straight trees are cloned to produce offspring that can be planted, cultivated, and harvested into sawlogs. This eliminates the cull trees ordinarily found in tree plantations from seedlings produced by conventional means (from seeds).

Many plant species have both male and female plants that are quite different. Usually, the female plants produce flowers and fruits that are used commercially. However, some plants, such as asparagus, are grown for the male plant. Cloning can greatly enhance the efficiency of propagation because only male or female plants can be cloned.

Methods of Plant Cloning

Remember from studying Chapter 4 that there are several very important differences between plant and animal cells. Plant cells have a large central vacuole, a cell wall, and organelles called plastids. Within the plant cell there are three types of **plastids**. These are chloroplasts that use solar energy to make carbohydrates, leucoplasts that provide storage, and chromoplasts that manufacture color in fruits. Most important for vegetative propagation of plants, each plant cell has the ability to produce an adult plant under specific environmental conditions. In order to generate an adult plant from a specialized plant cell, that cell must be able to dedifferentiate (divide into different types of specialized cells). Specialized animal cells do not have the ability to dedifferentiate and develop into a genetically identical adult animal, though reproductive cloning technology, such as nuclear transfer, may be used to achieve similar results (Chapter 7). Given differences in the ability of cells to dedifferentiate, cloning is a much easier process in plants than animals.

Remember from the chapter on animal cloning that identical twins are naturally occurring clones. Cloning also occurs naturally in plants but with a much greater frequency. There are many plants that naturally reproduce by asexual means and are natural clones (Figure 8–5). A producer may simply take advantage of this naturally occurring process by enhancing or helping the process along.

Figure 8–5
Many plants naturally reproduce by asexual means.

© Cengage Learning 2013

Cloning by Separation and Division

Two methods used for obtaining new plants from specialized plant parts are separation and division. In **separation**, the plant parts are merely pulled apart because the plant naturally separates the parts for the production of the new plants. In **division**, the producer cuts the plant part into sections and grows a new plant from each section.

Separation and division make use of specialized parts of the plants such as bulbs, corms, tubers, stolons, rhizomes, and crowns. Runners, or **stolons**, are specialized stems that grow on top of the ground and reach out horizontally. **Rhizomes** are also specialized underground stems that are used in propagation by division. They are generally larger than stolons and can be broken or cut apart and the parts used to establish a new plant. Irises are plants that are easily propagated by rhizomes (Figure 8–6).

Courtesy of ARS/USDA #K7461-11

Figure 8–6
Irises are among the plants that can be propagated by rhizomes.

Also, many grasses, such as Bermuda grass and Johnson grass, are propagated by rhizomes. An interesting characteristic of grasses of this type is that they also produce seeds. New plants from seeds are the result of sexual reproduction and combine the genes of two different plants. New plants generated from the rhizomes of the parent plants are genetically identical to the parent plants.

Sometimes a stolon or rhizome may have an enlargement called a **tuber** that can be cut apart and new plants grown from each section. For example, a potato is a tuber (Figure 8–7). On the surface of the tuber are several buds called eyes. Each of these buds is capable of sprouting a new plant when cut from the potato and planted. The resulting potato plants are exact genetic copies of the parental potato plant.

Many plants are propagated by means of **bulbs**. There are two kinds of bulbs: tunicate and nontunicate. **Tunicate** bulbs have dry outer layers of membranes that are the result of last year's growth. They are made up of layers of leaflike membranes

Figure 8–7
A potato is an example of a tuber that can be cut into sections for planting. Notice the sprouts growing from the potato "eyes."

and when cut into cross sections appear to have concentric rings. The onion is a good example of this type of bulb (Figure 8–8). Tunicate bulbs propagate naturally by growing small bulblets around the bulb. As these bulblets may each grow into a separate plant, growers periodically dig these bulbs and separate the bulblets in order to produce new plants. These bulbs may also be cut into segments. Each cut portion of the parent bulb is stored for 1–2 weeks. After that, they are planted, and they produce bulblets from the bulb cutting.

Nontunicate bulbs are also known as scaly-type bulbs because of the layers of scales on them. Each of the outer scales can be separated and planted. From each of these scales, a new plant will grow. Lilies are an example of a plant having a nontunicate, or scaly-type, bulb.

Corms are another type of underground stem used in propagation (Figure 8–9). Corms differ from bulbs in that they are solid and have nodes and inner nodes. A node is a place on a stem where a new leaf or other plant part emerges. Propagation takes place through the production of new corms (called cormlets) from the old corm. Growers separate the cormlets from the parental corm and plant them. Also, the corms may be divided into sections with each section producing new plants. Examples of plants that have corms are gladiolus and garlic.

Cloning by Cuttings

Producers may propagate plants by taking cuttings taken from parental plants with desirable traits. This method has been in use for quite a long time. One early example of vegetative propagation by cuttings is the story of a single apple tree found in Iowa in the late 1800s. Apparently, the apple tree was a mutant plant that came from a tree having very different characteristics. A mutant is an organism whose genetic makeup has been altered through a random change in the genetic code. These random changes, or mutations, are often the result of environmental sources of radiation, such as sunlight, or mistakes made by cellular DNA replication. Some mutations have no effect on the physical characteristics of an organism, but some mutations do result in new traits or characteristics. Several varieties of commercially produced plants have been derived from naturally occurring

© Linda Holt Ayriss/Photodisc/Getty Images

Figure 8–8
Onions are a good example of a tunicate bulb.

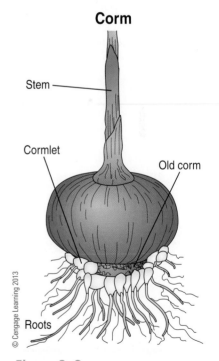

© Cengage Learning 2013

Figure 8–9
Another type of underground stem, called a corm, is used to propagate plants such as gladiolus and garlic.

Figure 8–10
One of our most popular varieties of apple, the Red Delicious, originated from a single tree.

mutants. In this example, the mutant apple tree had apples that were far sweeter than any known variety at the time. The tree produced a very delicious apple that could be eaten fresh. In fact, the apple tasted so good that the new variety was called Red Delicious. By taking cuttings from this tree, the new variety spread to all areas of the country and became one of the most popular varieties ever grown. Today, many thousands of acres of Red Delicious apples are grown in this country (Figure 8–10). Without early cloning technologies, such as the use of cuttings, this variety would have died out when the original tree died.

Depending on the type of plant, various parts of the plant, such as cuttings from leaves, stems, roots, or buds, may be used to grow new plants. Plants are completely different from most animals in that they can regenerate an entire plant from a severed part of the parent plant. Of course, most animals do not possess this ability. A severed part of an animal cannot produce a new animal, at least not with our present technology. The use of plant cuttings offers some advantages over other methods of vegetative propagation. With large trees and shrubs, a lot of plant material, including stems, leaves, and roots, is available for cuttings that may be used to reproduce the plants. Theoretically, any cell of the plant should be able to replicate itself because all the genetic coding for the entire plant is held within each cell. A large tree or shrub can provide a lot of stems, leaves, or roots from which to grow new plants. In addition, growers are able to control the vegetative propagation process because they do not generally have to wait on the plant to produce bulbs, tubers, corms, or other specialized reproductive organs to develop. However, in some plants, cuttings are best taken at certain times of the year. For example, hardwood cuttings grow best if they are taken during the dormant season (Figures 8–11A and B).

Environmental Conditions for Vegetative Propogation

In order for the plant to regenerate itself, the grower must create the proper conditions that will allow the development of roots and shoots from the plant cutting. As the cuttings do not have roots, the grower must create conditions that will retard the growth of leaves and stems until the roots can grow and develop to support the leaves and stems. These environmental conditions include the proper temperature, moisture, air movement, and light.

Figure 8–11A
Many cuttings can be taken from a tree or shrub.

Figure 8–11B
Long limbs are cut into short cuttings.

Proper temperature is important because photosynthesis influenced by air temperature. Generally, higher air temperatures promote higher rates of photosynthesis. For a plant cutting the grower wants to root, a high rate of photosynthesis may not be desirable. Photosynthesis encourages the use of stored plant energy for the production of leaves and buds. At a lower rate of photosynthesis, this energy can be used to promote the growth of roots on the cutting. Although the proper temperature may vary with the type of plant being propagated, the temperature is usually best

Figure 8–12
To propagate properly, the cuttings must be kept at the right temperature.

between 70°F and 80°F in the daylight hours and 60°F and 70°F during the night (Figure 8–12). Soil temperature should be approximately 5°F to 10°F higher than air temperature. Higher temperatures in the root area stimulate the oxidation of fatty acids. This creates a substance known as suberin, which promotes healing in the area where the cutting was severed.

A moist growing environment, both in the media surrounding the cutting and in the surrounding air, is required for cuttings to develop roots. Media is the semisolid material the plant cutting is placed in to root. The media must be kept moist but not wet because excess water will deprive new roots of oxygen and will promote decay of the cutting. Roots require oxygen in order to grow and thrive. Proper circulation of air through the media is important to allow oxygen to reach the part of the cutting that will grow the new roots. This means that the media must be of the proper consistency to allow the flow of air to the roots.

Relative humidity refers to the amount of moisture in the air surrounding the plant. This amount is expressed in terms of the percent of moisture the air will hold before condensation occurs. Leaves must retain their moisture in order to produce carbohydrates and plant hormones, or **auxins**, necessary for the production of roots and the rest of the plant systems. A high relative humidity prevents the plant cutting from losing too much moisture to surrounding air through transpiration, though too much moisture can promote the decay of the plant cutting. A range of 60–80 percent relative humidity is generally considered ideal for the rooting of most plant cuttings (Figure 8–13).

Figure 8–13
The cuttings must be kept moist.

© Cengage Learning 2013

Figure 8–14
Proper lighting is essential for propagating cuttings. Too much or too little can be detrimental.

© Cengage Learning 2013

Light is also important in producing new plants from cuttings. Remember that light is the energy used by the plant to undergo photosynthesis. The more light a plant receives, the greater the rate of photosynthesis. Although a certain amount of light is necessary for the cuttings to take root, too much light can be detrimental (Figure 8–14). The same problem is encountered with excessively high temperatures; the energy is put into the growth of leaves and not roots. To properly root, plant cuttings should be shielded from intense light.

"FasTracking" Plum Breeding

Plum, peach, cherry, apricot, and almond trees—what do they have in common? Well, they are all members of the genus *Prunus*, and their fruits are well loved among American consumers. But perhaps the most significant similarity is that the trees take a very long time to mature and produce fruit.

The time it takes for these fruit trees to reach maturity from seed has a considerable effect on the development of new varieties with desired traits, such as an improved mix of sugars or resistance to disease. "Fruit tree breeding still remains a slow, arduous process that has changed little over the centuries," explains horticulturist Ralph Scorza. "In addition to the long juvenile period (3–10 or more years during which trees do not fruit), other limitations include the need for large land areas with significant field costs and yearly limitations on flowering and fruiting related to chill and heat requirements. Temperate fruit tree crops require a period of dormancy to induce flowering and to bear fruit." But what if there were a way to get around those limitations?

That is exactly what Scorza and his colleagues at the USDA-ARS Appalachian Fruit Research Station in Kearneysville, West Virginia, aimed to do when they started "FasTrack," an advanced fruit tree breeding system. Plant physiologist Chinnathambi Srinivasan, molecular biologist Christopher Dardick, and geneticist Ann Callahan join Scorza on this project.

The team is focused on improving the breeding system for plum, a fruit the scientists have spent many years working on. With FasTrack, they have found a way to lessen the time it takes to create new, improved plums. Instead of taking 15–20 years to breed trees with a combination of desired traits, such as disease resistance and high-quality fruit, the scientists can now accomplish this task in just 3–5 years.

USDA/ARS, #D2142-1. Photo by Chinnathambi Srinivasan

Rather than the 3–10 years normally required for a seedling plum to produce fruit, FasTrack plum lines carrying the early-flowering gene produce fruit in less than a year after being planted from seed. **(D2142-1)**

Inducing Early Flowering

Genes for early flowering have been previously reported in several plants. By over-expressing certain flowering genes, scientists were able to induce early flowering in the model species *Arabidopsis* as well as in poplar and citrus. "But this early-flowering construct was never developed into a practical system for breeding," says Dardick. "Our research represents one of the first efforts to implement this technology in fruit trees."

Unlike conventional breeding, FasTrack uses plum lines that have been transformed with a special gene called "*PtFT1*" that induces early flowering. The scientists inserted this gene, previously discovered in California poplar (*Populus trichocarpa*), into the plum cultivar Bluebyrd to stimulate early and continual flowering.

The research team, spearheaded by the genetic engineering work of Srinivasan, produced 196 transgenic plum plants in total, many of which flowered and set fruits within 1 year. These plants do not look like trees, but rather like shrubs. And instead of single flowers, some buds produced flower clusters.

"Because of their growth habit, these transgenic trees are not suitable for standard orchard practices," explains Dardick. "So we keep them in the greenhouse, which provides a controlled environment where we can continually makes crosses all year long and produce offspring at a faster rate. By coupling FasTrack technology with new molecular marker technologies, we will ultimately be able to produce a new variety in less than half the time it would take with conventional breeding practices."

At the very last step of breeding in the greenhouse, the scientists select trees that do not have the *PtFT1* gene but still possess traits of interest, such as sweet fruit, resistance to a certain disease, or higher nutrient levels. In this way, FasTrack technology uses genetic modification during the breeding process, but in the final step, the trees that are selected for use as cultivars are not genetically modified; that is, they do not have the *PtFT1* gene, do flower normally, and are no different than if they had been produced through traditional breeding practices.

The scientists are the only group in the country conducting this kind of research at the moment. According to the researchers, FasTrack can be adapted for any plant, but it is more useful for plants that take a long time to mature.

The technology can also be used to produce fruit in new ways, especially considering the challenges we face in dealing with climate change. The transgenic tree's shrublike stature could make it possible to grow fruit in greenhouses or high tunnels. And because it does not require a chilling period, it could be adapted for growing in tropical climes and could be used to produce fruit year round. It might also be appealing to home gardeners as an ornamental plant that continually flowers and fruits.

—Agricultural Research *magazine, by Stephanie Yao, formerly with ARS.*

Courtesy of Dr. Frank Flanders

Figure 8–15
Cuttings are dipped into a plant rooting compound that contains hormones to stimulate root growth.

Growth Regulators

The reproduction of cells is controlled to a large degree by plant hormones called auxins. When roots begin to grow on plant cuttings, the process is stimulated by growth hormones. In certain plants, these hormones are released naturally, and propagation can occur readily using cuttings. Other plant cuttings require externally applied auxins. The cuttings are dipped in a powder containing the root growth-stimulating auxin or is soaked in a solution containing the hormone (Figure 8–15).

Layering

Some plants are difficult to root from cuttings. These plants can often be propagated by a process called **layering**. In layering, a portion of the plant is covered with soil or other material, and the growth of roots is stimulated while still attached to the parent plant. This method is used when the other methods do not work well. There are disadvantages to using layering. The cost of layering is more than that of many other methods because of the high cost of the manual labor required to complete the task. In addition, fewer new plants can be started by layering than with the use of cuttings.

The process of simple layering involves the cutting of a notch in a branch or stem of a plant. The wounded place (notched tissue) is then covered with soil with the tip of

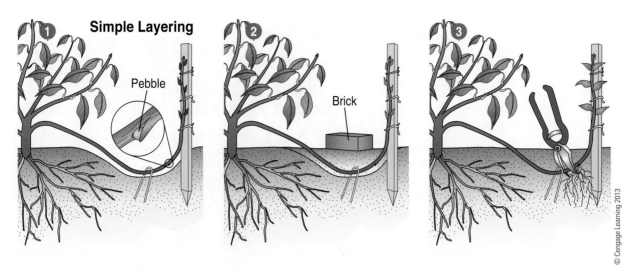

Figure 8–16

In simple layering, a notch is cut in a stem, and the stem is buried.

the branch or stem exposed above the soil line. The wound in the plant stimulates the growth of undifferentiated cells called callus cells. This means that they have not been genetically programmed to become a certain type of plant tissue, such as a leaf, root, or stem. Callus cells are capable of differentiating into root cells, and as new roots begin to grow at the site of wounding, a new plant develops (Figure 8–16). The new plant is cut free from the parent plant and planted elsewhere. Broadleaf evergreen trees and shrubs such as rhododendron and magnolia are propagated by this method.

In mound layering, the top portion of the parent plant is pruned back in the dormant season. In the late spring when new growth appears, soil is mounded up around the base of the plant, and part of the new growth is covered (Figure 8–17). By the end of the next dormant season, the shoots that were covered have developed new roots. These newly rooted sections are separated from the parent plant and planted in another place. Currants and gooseberries are propagated by mound layering.

Tip layering is a method by which the new growth at the tip of a parent plant is placed in the ground and covered with soil. The tip continues to grow downward in the soil. As growth continues, **meristem** cells in the tip develop into new roots and shoots (Figure 8–18). Once new shoots

Mound Layering

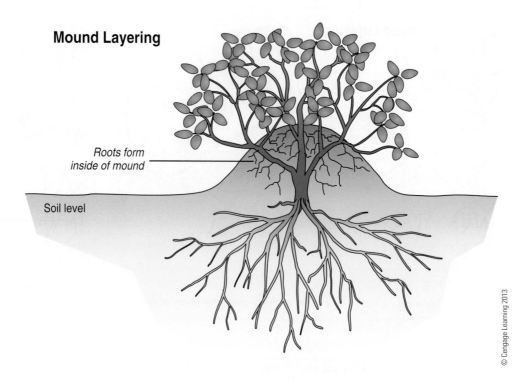

Roots form inside of mound

Soil level

© Cengage Learning 2013

Figure 8–17
Soil is mounded around the base of a plant in the process of mound layering.

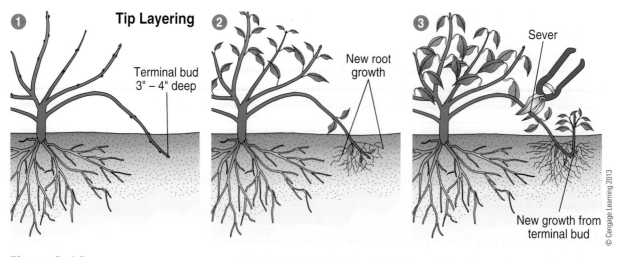

❶ Tip Layering

Terminal bud 3" – 4" deep

❷

New root growth

❸

Sever

New growth from terminal bud

© Cengage Learning 2013

Figure 8–18
Terminal buds are covered with soil to produce a new plant through tip layering.

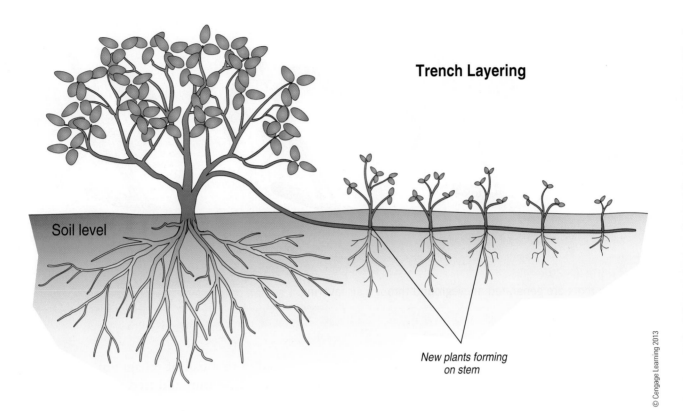

Trench Layering

Soil level

*New plants forming
on stem*

Figure 8–19
In the process of trench layering, a whole stem is buried, and several new plants are rooted.

emerge from the soil, they are removed and separated from the parent plant. Tip layering is used to reproduce plants such as blackberries, raspberries, and dewberries.

Trench layering is used in propagating vines, such as muscadine grapes and philodendrons. This method involves the covering of an entire stem in the soil. The nodes of the stem are notched before the stem is covered, and new roots are formed at the nodes where they were notched (Figure 8–19). Several new plants can be started from a single limb using this method. A variation of this method is called serpentine layering. In this variation, only the nodes are covered, and the rest of the stem is left uncovered.

Layering can also be done without covering plant parts with soil. This method is called air layering because the process is done in the air instead of in the soil. When propagating woody plants using this method, a healthy, fast-growing limb is selected, and a ring of bark is removed.

Air layerage

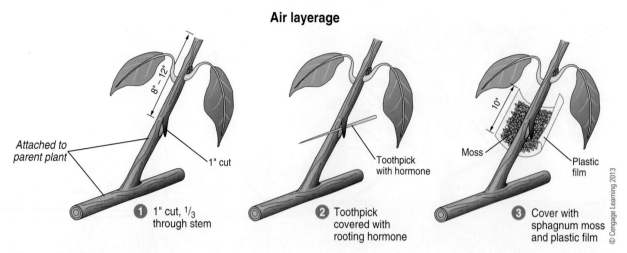

Attached to parent plant

8" – 12"

1" cut

❶ 1" cut, ⅓ through stem

Toothpick with hormone

❷ Toothpick covered with rooting hormone

Moss

10"

Plastic film

❸ Cover with sphagnum moss and plastic film

© Cengage Learning 2013

Figure 8–20
New roots are generated aboveground through air layering.

The wound is covered with a ball of sphagnum moss. The moss is covered with plastic film and tied to prevent the escape of moisture (Figure 8–20). After several weeks, new roots are formed at the place where the bark was removed. The branch is severed, and the new plants are planted. Herbaceous plants are air layered in a similar manner, except instead of removing a ring of bark, a notch is cut in the limb, and a toothpick is placed in the notch. The wound is then covered in sphagnum moss and plastic film in the same treatment used for woody plants. Air layering is used to propagate large houseplants such as rubber plants. Some citrus trees are also reproduced this way.

Cloning by Grafting

A method of cloning that has been used for many years is that of **grafting**. In this technique, plant material from two separate plants is joined together to form one plant. This method is used with plants such as almonds and apples, that are difficult to propagate by any other method (Figure 8–21). For example, a hybrid apple tree might be difficult to grow from seed because the tree from seed will be different from either of its parents. If a young seedling is grafted with the shoot portion of a desirable tree, the results can be a tree that

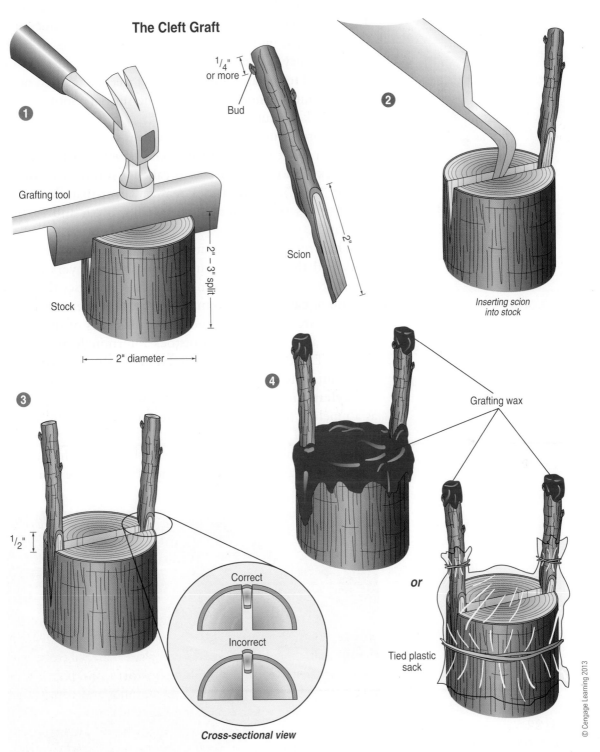

The Cleft Graft

1

Grafting tool

Stock

2" – 3" split

2" diameter

1/4" or more

Bud

Scion

2"

2

Inserting scion into stock

3

1/2"

Correct

Incorrect

Cross-sectional view

4

Grafting wax

or

Tied plastic sack

© Cengage Learning 2013

Figure 8–21

In grafting, two plants are joined together to form a new plant.

grows and produces like the desirable tree. This method is also used to produce specialty trees, such as dwarf trees and fruit trees, that will grow several different varieties of fruit. Also, trees may be grafted onto rootstock that is stronger or more suited to the area. English walnut trees are grafted onto black walnut rootstock because of the superior root system of the black walnut.

The procedure for grafting involves the cutting of two parts of a plant (usually a tree). The lower part is called a rootstock, and the upper part is called a **scion** (sometimes spelled "cion"). The rootstock may be larger than the scion, which gives the advantage of faster growth and earlier maturity.

The proper technique in grafting is to align the rootstock and the scion so that the **cambium** layers match (Figure 8–22). The wounding from the cut causes the production of **callus cells**. Remember that callus cells are those cells that are undifferentiated. The callus cells of the scion and the callus cells of the rootstock intermingle and begin to grow. These cells combine and differentiate to form new cambium cells. The new cambium cells then develop into the **xylem** and **phloem** vascular tissues. Xylem functions as a pipeline for transporting water and minerals from the roots to aerial portions of the plant. Phloem carries nutrients derived from photosynthesis down to the roots. When grafting, it is critical to make a connection between the xylem of

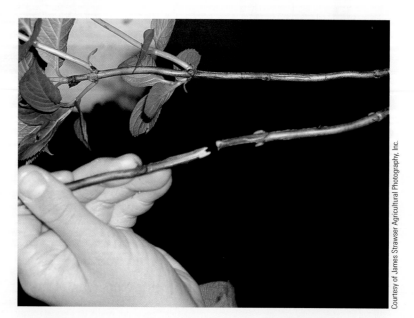

Figure 8–22

The proper technique is to join the rootstock and the upper part, called a scion.

the rootstock and the xylem of the scion. Likewise, a continuous connection of phloem is required for growth and development of the grafted plant.

Many types of grafting have been developed to fit specific purposes and specific plants. Grafting may involve using stems, roots, or buds. All the grafting methods rely on the same basic principle: make a healthy, functional connection between vascular tissues of the rootstock and scion. To ensure a successful graft, the rootstock and scion must also be bound tightly together and covered with wax or other material to prevent the plant tissues from drying out.

Cloning by Tissue Culture

One of the newer technologies involved in plant propagation is **tissue culture**. This process began in France in the mid-1960s when a French scientist named George Morel was attempting to develop orchids that would be disease free. He discovered that a very tiny part of the plant could be used to grow a completely new plant. Since that time, the process has developed into commercial use in propagating plants that are totally disease free.

Tissue culture is performed in a sterile environment, usually in a laboratory. A small amount of tissue is removed from a parental plant in order to regenerate a new plant (Figure 8–23). There are two big advantages to using this procedure. The first is that a large number of genetically identical plants or clones may be regenerated from a single parent. The second advantage is that the cloned plants are free of disease-causing organisms.

Plant growth and development takes place through cell division of undifferentiated meristem cells (Figure 8–24). Resulting daughter cells then differentiate and take on specialized functions. Some examples of specialized plant cells are mesophyll cells, found in leaves and the site of photosynthesis; epidermal cells, which makeup the protective outer layer of the plant; and root cap cells, which form a protective barrier at the tip of the root. There are many other types of specialized plant cells, and all are derived from unspecialized meristem cells found at the growing root and shoot tips of the plant.

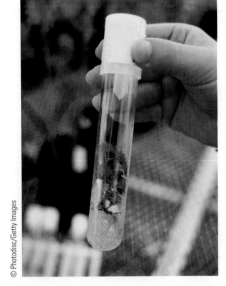

© Photodisc/Getty Images

Figure 8–23
Tissue culture involves the use of a tiny amount of plant tissue that is begun in a test tube or Petri dish.

© Cengage Learning 2013

Figure 8–24
Plant growth takes place through the multiplication of meristem cells. This is a meristem.

This means that, theoretically, each cell should be capable of growing a new plant. Growth takes place through the multiplication of meristem cells. The genes in each cell are programmed to differentiate. This means that some cells become stem cells, some become leaf cells, some become root cells, and so on. In meristem tissue, the cells are still undifferentiated. In the process of tissue culture, these cells are cut away from the plant before they are programmed to differentiate. By placing the cells in an environment that causes them to grow, they multiply and differentiate into cells, such as roots, stems, and leaves, that the plant must have to live. **Cell differentiation** may be influenced by internal biological signals, such as plant hormones (e.g., auxin), availability of nutrients or minerals, and the location of a cell within the plant. External environmental factors also influence cell differentiation, such as light, soil water, and air temperature. There are probably several other factors that influence cell differentiation. Scientists still do not fully understand the process. In order for tissue culture to work, all the factors mentioned have to be taken into consideration.

Tissue culture relies on the dedifferentiation of plant tissues into cells called callus cells. Ultimately, these callus cells will be given the appropriate plant hormones and nutrients to encourage growth of new roots and shoots. The first step in the process is the collection of tissue from a donor or parent plant. This tissue is called the **explant**. Care is taken to get the proper amount of tissue from the correct place.

The explant is thoroughly cleaned and sterilized. It is then placed in a sterile container, such as a Petri dish, containing a media composed of sterile minerals, nutrients, and plant hormones. As nutrient-containing media is an ideal environment for the growth of microorganisms, the explant, container, and solution must be maintained in a sterile environment.

Working in a sterile in vitro ("within the glass") system means that the plants grown in tissue culture will be disease free, which is beneficial to commercial producers. Callus cells are later transferred to media with the plant hormones and nutrients that will promote the growth of new roots and shoots. Once the regenerated plantlets have leaves and roots, they may be transferred to soil or maintained in vitro.

As more research is conducted and new methods of conducting tissue culture are developed, this type of propagation is likely to become even more common. Today, several large enterprises use tissue culture as the basis for their propagation. One of the primary reasons is that tissue culture allows the shipment of plant material between different countries without the fear of spreading plant diseases. The future holds promise for many applications of this technique.

Plant Research and Vegetative Propagation

Many plant research laboratories use tissue culture to generate cloned populations of plants in order to study genetically identical plants under different environmental conditions (Figure 8–25). It is preferable to work with plants having the same genetic background when conducting these types of experiments in order to more clearly determine the influence that a particular environmental treatment or stress has on plant growth. Results may be applied to developing

Figure 8–25
Tissue culture is a very promising technique for plant propagation.

hardy varieties of plants for use in agricultural systems, where severe weather, plant diseases, and insect pests often have a negative impact on plant growth.

As more fully explained in Chapter 9, biotechnology has been used to successfully develop crops with specific useful traits, such as insect resistance and herbicide tolerance. As a critical part of the procedure for making biotech crops, tissue culture is used to regenerate genetically modified plant cells. In the future, these plants will completely revolutionize the agricultural industry by creating superplants that will be more productive and also yield products far more valuable than today's conventionally grown plants. The propagation of these plants may very well be done by cloning because if a superplant is genetically engineered, the genetic integrity of the plant must be kept intact. As propagation through seed means that genes from a different plant are involved, there is a danger that the gene pool may be diluted. Cloning can solve the problem and keep the characteristics of the plant that make it "super."

Summary

Plant cloning has been practiced by producers for centuries. Early forms of plant cloning include the vegetative propagation of genetically identical plants by corms, bulbs, and tubers through separation and division. Cutting, layering, and grafting have also been traditionally used to asexually reproduce plants with desirable traits. Cloning plants are an advantage to growers because they can be sure that the new plants propagated will be genetically identical to the superior plants selected as parents.

Beginning in the 1960s, commercial growers and research laboratories have increasingly used tissue culture to generate large populations of plant clones. Tissue culture relies on the ability of specialized or adult plant cells and/or explant tissues to dedifferentiate into callus cells that may, in turn, be stimulated to develop into new plants.

CHAPTER REVIEW

Student Learning Activities

1. Visit a local nursery and determine how many of the plants sold there are cloned by asexual reproduction. Make a list of all the steps that go into the process.

2. Locate a tree, shrub, or plant that you think is an exceptional plant. List the characteristics that you feel make it superior. Determine the feasibility of cloning the plant. How is that species of plant currently propagated?

3. Bring a plant to class that is propagated by cloning. Report to the class on the desirable characteristics of the plant. Also describe how the plant is cloned.

4. Compile a list of the plant characteristics that make them easy to clone. For example, what characteristics do commonly cloned plants share?

Fill in the Blanks

1. Many plants reproduce by means of _____ through a process known as _____ _____ or _____ _____.

2. To produce _____, plants must complete an entire cycle of _____ production, _____, seed _____, and _____.

3. Runners, or _____, are specialized _____ that grow on top of the ground and reach out _____.

4. New plants from _____ are the result of _____ _____ and combine the _____ of two _____ plants.

5. Producers may _____ plants by taking _____ taken from _____ plants with _____ traits.

6. A _____ is an organism whose genetic makeup has been _____ through a _____ _____ in the genetic code.

7. When _____ begin to grow on plant _____, the process is stimulated by _____ hormones called _____.

8. The procedure for _____ involves the _____ of two parts of a plant. The lower part is called a _____, and the upper part is called a _____.

9. The first step in the process of _____ _____ is the collection of tissue from a parent or _____ plant. This tissue is called _____.

10. Working in a sterile _____ _____ ("within the glass") system means that the plants grown in _____ _____ will be _____ _____, which is beneficial to commercial producers.

True or False

1. Plant cloning is one of the newest forms of biotechnology.

2. Plants that have seedless fruit must be reproduced through vegetative propagation.

3. There are male and female plants that have different physical characteristics.

4. All plant cells have the ability to create an adult plant under the correct conditions.

5. Cloning plants are much harder than cloning animals.

6. A tuber is a specialized root.

7. Corms differ from bulbs in that they are solid and have nodes and inner nodes.

8. Cuttings are usually only taken from leaves.

9. Plants that are not easily rooted from cuttings can often be propagated through layering.

10. One of the disadvantages of tissue culture is that many genetically identical plants can be produced in a short period of time.

Discussion

1. List at least three advantages to using plant cloning.

2. Why is natural cloning in plants much more common than natural cloning in animals?

3. Distinguish between separation and division.

4. Compare and contrast tunicate bulbs and nontunicate bulbs.

5. Describe the conditions necessary for a plant cutting to properly root.

6. What is a callus cell?

7. List and briefly describe three types of layering.

8. Explain the difference between xylem and phloem.

9. Describe the process of grafting.

10. Why is it important to clone plants that have been genetically altered?

CHAPTER 9

Biotechnology in Plant Science

OBJECTIVES

When you have finished studying this chapter, you should be able to:

- Explain why insect control is so important in plant production.
- Describe how bacteria are used to create insect resistance in plants.
- List the advantages of plants that have been genetically modified to make them insect resistant.
- Explain how *Bacillus thuringiensis* (Bt) kills insects and remains harmless to humans.
- List crops that have been engineered to contain the Bt gene.
- Explain how Bt crops help alleviate mycotoxins.
- Explain the reason for making crops herbicide resistant.
- Describe the problems caused by weeds in crops.
- Describe how plants are genetically engineered to prevent viral infections.
- Describe how crops can be made tolerant of climatic conditions through genetic engineering.
- Describe how crops are being engineered for biofuels production and plant-made products.
- List the ways in which biotechnology, such as genetic engineering, is currently being used in producing plants.

Introduction

For many thousands of years, humans have selected useful traits in plants grown for food, shelter, and clothing. The plants we use in our everyday lives are dramatically different than those that are found in the wild (Figure 9–1).

(A)

(B)

Figure 9–1
The plants and crops we grow are very different from those that grow in the wild. For example, carrots (A) were developed from a plant known as Queen Anne's lace (B).

The process of **selective breeding** to produce edible, farmable plants is directly responsible for the abundant food supply we enjoy today. Although selective breeding has been successful over thousands of years, it is slow and painstaking compared to new methods of plant improvement through genetic engineering and other forms of plant biotechnology. This chapter deals with some of the changes brought about by genetic engineering that are already in use.

Insect Control through Biotechnology

Over the thousands of years that humans have been growing plants for their own use, one of the greatest problems has been that of insects devouring much of the crop. Until relatively recently, common practice was to plant twice as much as needed because insects will get half the crop. In fact, even with the advent of today's highly effective **pesticides** and management techniques, it is estimated that about 40 percent of the world's food supply is lost to pests, diseases, and spoilage (Figure 9–2).

Sterile Male Insects

People have known for many years that a very effective way of controlling insects is to interrupt their life cycle. This can be accomplished in a number of ways, all involving

© Angel Simon/www.Shutterstock.com

Figure 9–2
It is estimated that 40 percent of the world's food supply is lost to pests, disease, and spoilage.

Figure 9–3
The release of thousands of sterile male insects in infected areas is an effective way of eradicating the insect.

their reproduction. A very effective way, using biotechnology, is to release thousands of sterile male insects over the infected area. In a laboratory, male insects are treated with radiation or other treatments to render them infertile. The process allows the males to still be able to mate. The wild females mate with the sterile males, and no offspring are produced. Through a period of several years of saturating the area with sterile males, an insect pest can be eradicated (Figure 9–3).

One of the most successful uses of this method was that of eradicating the boll weevil in the southeastern United States. For much of the twentieth century, the boll weevil devastated cotton crops all across the South. Each year, millions of dollars were spent on pesticides to kill this insect. In the latter part of the twentieth century, a concentrated effort was made to saturate the entire South with sterile male boll weevils. The method was so successful that the boll weevil has for all practical purposes been eradicated from the South (Figure 9–4). Other insects, such as the screwworm, have also been eliminated as a serious pest. Not only does this save huge amounts of money each year, but the environment is kept cleaner by using fewer pesticides.

Courtesy of ARS/USDA Image #K2742-6

Figure 9–4
Through the use of biotechnology, the boll weevil has been eliminated as a serious pest.

Insect-Resistant Plants

Perhaps the greatest impact that biotechnology has had on plants occurred with the development of **genetically altered crops** that have a built-in resistance to insects. Not only has this innovation saved producers millions of dollars in pesticides, but the environment has also been made cleaner since fewer chemicals are used in crop protection. Labor costs, fuel expenses, and machinery operating costs are greatly reduced when plants have a built-in resistance to insect damage. These economic factors have a great impact on modern producers in the United States and other developed countries, but the most dramatic effects may be in Third World countries. Here, producers are most often not able to purchase the pesticides necessary for plant protection. Even those few producers who may have the financial means to buy pesticides may not have a readily available source for the chemicals. If plants were available that provide a built-in protection, the purchase of pesticides would be unnecessary (Figure 9–5.)

The insect resistance came about as a result of the development of transgenic plants. Remember from a previous chapter that transgenic organisms have genetic materials that were inserted from a different species. In this case, the genetic material came from an entirely different kingdom of organisms.

Courtesy of ARS/USDA #D667-1. Photo by Peggy Greb

Figure 9–5
These scientists are examining wheat plants that are resistant and susceptible to Hessian fly. Both plants are the same age, but the one on the left has had its growth stunted by Hessian flies.

Figure 9–6
Bt spray has been used for many years as an organic pesticide.

Figure 9–7
Bacillus thuringiensis (Bt) secretes a toxin that disrupts the digestive system of insects. After only a few bites of peanut leaves with built-in Bt protection, this lesser cornstalk borer larva crawled off the leaf and died.

Scientists have known for many years that the bacteria ***Bacillus thuringiensis*** (commonly known as **Bt**) was effective in killing insects (Figure 9–6). This soil-borne bacteria was first discovered in 1911 when scientists realized that the toxins secreted by these microorganisms disrupt the digestive process of insects. It was further discovered that these toxins were nontoxic to mammals and other warm-blooded animals such as birds.

Bt has been used for many years as an organic means of controlling insects (Figure 9–7). There are over 280 different strains of the bacteria, and many of the strains produce different toxins. The different toxins affect different insects.

For example, some toxins produced by Bt are toxic to caterpillars, others may be toxic to beetles, and others may be toxic to flies. The spray is still available to producers and gardeners who believe that the use of organic means of pest protection is safer and more desirable. The biggest drawback of Bt sprays is that they break down very rapidly when exposed to sunlight. While organic gardeners see this as an advantage, large producers argue that the pesticide may break down during the application process and be rendered ineffective. In fact, any applied pesticide will be effective for only a short amount of time. Once an applied pesticide has dissipated, crop plants are once again vulnerable to insect damage.

The problem of pesticide breakdown after application was partially solved by creating plants that manufacture Bt toxins. First, scientists studied the proteins responsible for Bt activity to make sure they were safe to use in biotech crops. These proteins, called Cry1A proteins, were isolated from Bt bacteria. Studies showed that the proteins are toxic to insects because they attach to cells in the digestive tract of target insects and cause internal bleeding. Similar toxic reactions do not occur in mammals, including humans, because their digestive tracts are acidic. Bt toxins need an alkaline environment, like that found in the insect digestive tract, in order to work.

Currently, there are millions of acres of Bt crops produced all over the world, and the amount is increasing. Soybeans, corn (maize), potatoes, tomatoes, apples, rice, and several vegetable crops have all been produced using the Bt gene. The potential exists to place the Bt genes into any crops that would benefit from the insect resistance technology. The majority of soybean and maize produced in the world have been genetically modified for insect resistance. These are staple crops widely used in food production, and consumers in the United States have been safely eating Bt food products for over 10 years. In Europe and elsewhere, anti-biotech activists and some consumers still believe that biotech crops may not be safe to eat, though their reservations are not supported by science. Consumers still will not accept genetically altered potatoes (Figure 9–8). Producers who have an opportunity to try biotech crops often see a great financial benefit and adopt this new technology, but other members of the community are slower to appreciate the benefits. Consumer concerns are discussed in detail in later chapters.

© Catlook/www.Shutterstock.com

Figure 9–8
The public would not accept genetically altered potatoes.

Another benefit of Bt-mediated insect resistance when compared to spraying pesticide is that all plant parts are protected. One of the major problems faced by producers who use chemical pesticides is getting the chemicals to where the insect pest is invading the crop plant. If the insect is eating the leaves of the plant, spray can be applied over the top of the plant to cover the leaves. Even this can be a problem if the foliage of the plant is so thick that the spray cannot reach all the leaves. Insects may attack the roots of the plants, and this makes pesticide coverage difficult. Likewise, an insect may burrow into the stem of the plant, and a topical application of chemicals cannot reach the pest.

Certain **pesticides** called **systemics** are taken into the plant and may be able to get at the insect pest, but there are three problems with systemics. First, the plant may not take up enough of the insecticide to be effective. Second, systemic pesticides can be a problem with food crops since the pesticide is taken into the plant system and may leave a residue in the edible parts of the plants. Long **withdrawal periods** may be required before produce from plants treated with systemic pesticides may be harvested and processed for consumption. (A withdrawal period is the time, from the application of the pesticide until harvest, required to render the produce safe to eat.) Third, systemic pesticides are expensive.

One of the most dramatic examples of the use of Bt insect resistance has been in maize (corn). One insect pest that causes the greatest damage to corn crops is the European corn borer. This insect is a very serious problem in most of the corn-growing regions of the country. Corn borer is difficult to control because the larvae burrow into the cornstalk and cause severe damage to the plant (Figure 9–9). Because the pest is hidden and protected by being on the inside of the plant, topically applied sprays cannot reach it once it has begun the damage. Corn plants carrying the Bt gene are toxic to the corn borer larvae, and this toxicity protects the entire plant. Other strains of the Bt gene can aid in the control of corn insects, such as armyworm and corn earworm. Although the effects are not as dramatic as those achieved in controlling the European corn borer, the use of the Bt gene allows for lesser amounts of pesticide use.

An added benefit of the use of Bt crops, such as soybeans and corn, is a significant reduction of toxins called **mycotoxins**. These very potent poisons are produced by a fungus

Figure 9–9
The use of Bt corn greatly reduced damage from the European corn borer.

that grows on grains or other feedstuffs. The use of the Bt gene helps in two ways. First, there is less insect damage to the crop, which limits where the fungus may infect the plant. Fungi often enter a plant where there is an opening caused by a feeding insect. As the fungus grows, spores are released that give off poisons called mycotoxins. Second, the Bt toxins actually kill the fungus that produces the mycotoxins. In addition, stored feeds, such as grain and silage, have less growth of mold and fungi if they are produced from genetically modified plants containing the Bt gene.

Herbicide-Resistant Crops

Another serious problem faced by producers are weeds that compete with crops for soil nutrients and space. Weeds are considered agricultural pests in addition to insects and disease organisms. Each year, millions of dollars are spent in the attempt to lessen the negative effect of weeds on agricultural production. Weeds damage agricultural plants and animals in several different ways.

Weeds compete with crops for materials essential to growth (Figure 9–10). If allowed to grow unchecked, weeds get a large share of the soil nutrients that would otherwise go to the crop, thereby wasting these valuable nutrients. Also, weeds can compete with the crop for space and sunlight. Weeds that grow taller than the crop will shade the plants

Figure 9–10
Weeds are a serious problem because they compete with crops for nutrients, water, and sunlight. Note the weeds in this field.

and seriously reduce the amount of sunlight the crop plant can use for photosynthesis. This causes the crop plant to grow more slowly and to produce less.

Weeds may harbor insects and disease organisms that may spread to the crops. Many weed species attract pests that are harmful to the crops. With weeds growing among the crop plants, control of insects and diseases can be more difficult since the producer must deal with more than one type of plant. Likewise, disease organisms can be spread to crop plants through weeds that serve as hosts to disease organisms.

Weeds cause impurities in agricultural products. Soybeans that contain weed seeds are quite a bit less valuable than pure soybeans. Weed seeds can be poisonous to livestock and may be so close to the size and shape of soybeans that it is difficult to sort them out. Cotton that is picked among weeds will have trash in the lint. This reduces profit because the trash has to be removed from the lint.

For many years, humans battled weeds using only mechanical means. A hoe and a plow were about the only means available for controlling weeds in crops. This method not only was expensive but also could lead to soil erosion (Figure 9–11). Later, chemical **herbicides** were developed that were very effective in killing weeds. If weeds are to be removed from a crop, the chemical must do little or no harm to the crop. It would do little good to kill the weeds if the crop plants were also killed.

Some **herbicides** are called **nonselective** or knock-down herbicides. These chemicals kill all plants they are applied to and are most often used where all vegetation is to be killed. Other **herbicides** are **selective** and only kill certain types of plants. It sometimes seems almost magical that a crop full of weeds can be sprayed, and only the weeds are killed. Although these chemicals are effective in controlling weeds, they present problems. Any application of pesticides can be expensive. Machinery, fuel, and labor costs can add up to a large part of the production cost on many crops. The problem is compounded by the wide variety of weeds that infect crops. Selective herbicides are designed to kill certain weeds or types of weeds (Figure 9–12). For example, chemicals that kill only broadleaf plants will not kill grasses, and herbicides designed to kill grasses may not kill broadleaf plants. This presents a problem where fields are infested with both broadleaf and grass weeds (as mentioned previously, nonselective herbicides kill all types of plants—including the crop plant).

© rhambley. Image from BigStockPhoto.com.

Figure 9–11
Mechanical means of destroying weeds is expensive and can lead to soil erosion. Note the soil blown into the air.

Courtesy of ARS/USDA #K8042-1. Photo by Doug Buhler

Figure 9–12
Selective herbicides kill only certain plants. The corn on the left was treated with an herbicide. The corn on the right was not.

Gender Evolution in Strawberries: The Doorway to Enhanced Productivity

It is no surprise that strawberries are the most popular type of berry fruit in the world. This sweet, juicy, refreshing fruit is the perfect snack or end to any meal, especially on a hot summer day. But have you ever wished you could eat locally grown strawberries all year round? Well, new Agricultural Research Service research brings us one step closer.

Kim Lewers, a plant geneticist with the ARS Genetic Improvement of Fruits and Vegetables Laboratory in Beltsville, Maryland, and plant evolutionary ecologist Tia-Lynn Ashman, a colleague from the University of Pittsburgh's School of Arts and Sciences, discovered a model system for studying sex chromosomes in plants. This discovery opens up new opportunities for developing novel strawberry cultivars with increased fruit yields all year long.

Lewers and Ashman set out to determine the genetic control of reproductive dysfunction in strawberries because of its importance to fruit yield and quality. Reproductive dysfunction occurs when strawberry flowers do not bear fruit or do not produce enough pollen, resulting in small, misshapen fruit.

Strawberry breeders have long believed that strawberry plants can have one of three reproductive functions: male, female, or hermaphrodite. Male plants bear flowers that produce pollen but cannot set fruit. Female plants produce fruit if their flowers are pollinated, but cannot produce their own pollen. Hermaphrodites contain both male and female functions that enable them to flower, self-pollinate, and bear fruit. According to Lewers, strawberry farmers prefer their plants to be hermaphroditic so they do not have to plant more than one kind of strawberry in the same field in order to have any fruit to harvest.

Plants that have no pollen and cannot produce fruit, even when pollinated, are neuter.

Many breeders also follow the theory that genetic control of gender in strawberry plants is determined by one gene, and that there are three forms of the gene—scientifically referred to as "alleles"—at a single location on the chromosome that determines a plant's gender. According to this theory, the female allele is dominant, the hermaphrodite is semidominant, and the male allele is recessive.

Researcher Ashman collected plants of the wild strawberry *Fragaria virginiana* and crossbred them to create 200 offspring plants in hopes of better understanding the inheritance of dysfunction. "As our research progressed," says Lewers, "I began to wonder, 'What if dysfunction is determined by two separate loci, the places on a chromosome where a specific gene is located?'"

A New Gender Class

In order to answer her question, Lewers had to examine the data piece by piece. The team first inspected the offspring, giving each plant a score based on its "maleness" or "femaleness." Males were scored "male fertile" if they produced plump, yellow, pollen-filled anthers; those that did not have pollen were scored "male sterile." Similarly, plants were scored "female fertile" when at least 5 percent of their flowers set fruit, while "female sterile" plants bore less than 5 percent fruit. Plants with strong male and female traits physically demonstrate the dominant sex alleles in their DNA.

The next step involved mapping the genes that control reproductive dysfunction in *F. virginiana*. Genetic mapping is a process by which geneticists determine which genes are next to each other and, therefore, are usually

Plants without pollen that can set fruit if pollinated by another plant are female.

inherited together. The closer together the genes are, the higher the chances of their being inherited together. The process can include physical traits, like reproductive dysfunction, and molecular markers—tools geneticists use as DNA place marks or reference points.

Lewers, Ashman, and Ashman's postdoctoral fellow Rachel Spigler were able to create the first reproducible molecular-marker map of an octoploid strawberry. Most strawberries sold in grocery stores are octoploids—meaning in their natural evolution, the chromosomes have doubled and then doubled again to produce their current genetic makeup. The map the researchers produced can be used by strawberry breeders to help them naturally breed strawberries with better traits, such as disease resistance and year-round fruiting.

Two Genes Are Better Than One

Lewers also found that gender in strawberries is determined by two genes instead of one, and that the different alleles of the genes tend to be inherited together or passed to the offspring in pairs as they exist in the parents; this means they are physically next to each other on the chromosome.

The DNA map of the offspring shows that recombination—a process where chromosomes cross over and produce combinations of genes not found in the parents—occurs. This results in the presence of neuters, a gender class not taken into account by the conventional theory. Neither male nor female in function, neuters occur when the alleles containing male sterility and female sterility combine. Neuters physically look like females in that they flower and do not possess pollen; however, they do not produce fruit when pollinated.

"What we found is really quite extraordinary," says Lewers. "Before, neuters were not thought of as a possible gender class in strawberry. Our discovery of neuters shows that two loci control gender expression, not one, which means that this strawberry represents a very early stage in the evolution of chromosomes controlling gender in all plants."

Lewers's and Ashman's findings show that the gender determination in strawberries is influenced by two genes with different alleles of each gene on the chromosome. The presence of neuters in the offspring confirms that the two genes can recombine, a key step that has never before been addressed by the traditional theory on strawberry reproduction.

—By Stephanie Yao, Agricultural Research Service Information staff.

This problem has been partially solved by means of genetic engineering. Several years ago, a corporation called Monsanto developed a nonselective herbicide they called Roundup. This herbicide is very effective at killing a wide variety of unwanted vegetation, is relatively inexpensive, and is environmentally safe. It uses a substance called **glyphosate**, which works by stopping the action of an enzyme that serves a vital role in the production of certain amino acids that are essential to plants. The treated plant cannot produce the needed proteins and dies. Animals, humans included, are not affected by glyphosate. In fact, the Environmental Protection Agency (EPA) has rated Roundup as "essentially nontoxic" (Figure 9–13).

Scientists have been successful in genetically engineering corn, cotton, canola, and soybean plants that are tolerant of Roundup and similar herbicides. They accomplished this by inserting genetic material that allows other means of producing the amino acids blocked by glyphosate. The big advantage is that one herbicide (glyphosate) that is relatively cheap and safe to use can control all the weeds in a particular crop.

Figure 9–13

The EPA has rated glyphosate "essentially nontoxic." This scientist is preparing carrot samples to test for residue of the herbicide.

Disease-Resistant Crops

Plant diseases can be devastating to crops. Humans have dealt with these diseases for hundreds of years. In fact, history has even been changed as a result of plant pathogens. A prime example is a disaster that happened in Ireland during the mid-1800s. A main staple of the Irish diet was potatoes. The climate of the country was right for growing the crop, and many people depended on potatoes for their food. In the 1840s, a disease called potato blight almost wiped out all of Ireland's potato crop (Figure 9–14). As a result, hundreds of thousands of people starved to death, and over a million others left Ireland. Countries such as the United States, Canada, and Australia benefited from these immigrants. Many of the Irish immigrants were bright, talented people who made many contributions to their newly adopted countries. Another result was that disease-resistant varieties of potatoes were developed as a way of combating the disease.

Potatoes are by no means the only crop affected by diseases. Almost all crops, from grain to citrus fruits, are susceptible to diseases caused by fungi, viruses, or bacteria. Each year,

Courtesy of ARS/USDA Image #K5455-7. Photo by Scott Bauer

Figure 9–14
In the 1840s, a disease called potato blight caused widespread starvation in Ireland.

Courtesy of ARS/USDA #K8891-19. Photo by Scott Bauer

Figure 9–15
One environmentally friendly alternative to pesticides is genetic engineering of disease-resistant plants. These transgenic plums contain a gene that makes them highly resistant to plum pox virus.

many tons of pesticides designed to prevent plant diseases are applied to most of the crops we grow. One environmentally friendly alternative to pesticides is genetic engineering of disease-resistant plants (Figure 9–15). In the case of papayas and squash that have been engineered to resist viral infections, a gene for the viral protein coat was placed in the plant genome. The resulting genetically engineered plants are immune to viral infection, and this strategy is referred to as *pathogen-derived resistance*.

Climatic Tolerance

Many areas of the world have harsh climatic conditions that create problems for crop production. There may be insufficient rainfall, or the average temperature may be too hot or too cold for ideal growing conditions. Often, there are populations of people living in these areas who could benefit if the food-producing capabilities of the area were increased. Obviously, there is very little that can be done about the climate of an area. This means that the types of crops grown must be adaptable to the region.

In the past, a lot of progress was made in the breeding of crops designed for particular climates. During the 1970s, the **Green Revolution** included the development of crop varieties that could be grown in specific climatic regions (Figure 9–16). Wheat, corn, rice, and many other crop varieties were developed that could withstand drought conditions and a range of other conditions, such as cold temperature or saline soil. Varieties of fruit trees were bred for production in regions where they had never been grown before. Indeed,

Courtesy of ARS/USDA #D1560-1. Photo by Paxton Payton

Figure 9–16
This variety of peanuts was developed to be drought tolerant. Even with a deficit of water, the variety yielded fairly well.

© John A. Rizzo/Photodisc/Getty Images

Figure 9–17
Researchers have isolated genes that make plants drought resistant. This could lead to the development of plants that can be grown in harsh climates.

the Green Revolution made a lot of progress in food and fiber production, which has supported population growth over the last 40 years. However, if food, feed, and fiber production is to keep up with current population growth, scientists will have to improve crop yields much more.

A large step in the right direction is being taken through the use of genetically modified plants. Even in some of the harshest climates in the world, native plants grow and thrive. These hardy plants have developed genetic tolerance of environmentally stressful conditions, such as drought, heat, cold, saline, or nutrient-poor soil. Researchers have been able to isolate genes that make plants drought resistant (Figure 9–17). In some cases, **stress tolerance** genes have been successfully engineered into crop plants in order to enhance plant growth and yields.

Many thousands of acres of arable land are lost each year because of increasing soil salinity, which may be caused by buildup of salts from poor irrigation and drainage. There are also many acres that have never been farmed because of naturally high salinity from underground saline aquifers and lack of rainfall. By studying the biology of salt marsh plants and those that are "salt loving," or **halophytes**, researchers have identified genes that help salt-sensitive crops survive highly saline soil conditions (Figure 9–18). These genes allow plants to take up and store excess salts in plant cell vacuoles or tissues around the xylem, maintaining osmotic (water) balance and membrane potentials of the cells. Tomatoes, rice, wheat, and a number of other important crops are in commercial development.

In many developing countries, lack of irrigation systems and unpredictable rainfall make drought-tolerant, or **water use efficiency (WUE)**, crops a major benefit. Monsanto is currently testing water-use-efficient maize in Africa, and a number of similar public and private efforts are underway. Conversely, flood tolerant varieties of rice have been engineered for regions of Southeast Asia that are prone to flooding. Other crops in development have a **nitrogen use efficiency (NUE)** trait and require less chemical fertilizer, which is cheaper for farmers and better for the environment. Ultimately, many biotech crops on the market will have some combination of useful genetic traits. Crops with more than one genetically engineered trait are said to

© Pete Pahham. Image from BigStockPhoto.com.

Figure 9–18
By studying salt marsh plants such as this, scientists have identified genes that help salt-sensitive plants survive high-salinity soils.

have **stacked traits**. Monsanto and DowAgroscience have worked together to develop a varieties of soybean, maize, and cotton called SmartStax. These crops have been engineered with insect resistance, herbicide tolerance, and water use efficiency.

Other Uses of Plant Biotechnology

With the ability to custom-design plants through genetic engineering, the possibilities to produce food, feed, fiber, plant-made products, and biomass for renewable energy is limited only by our imaginations and our understanding of plant biology. Traits that increase crop yields or make crops less expensive to grow are called **agronomic traits**. Insect or disease resistance, herbicide tolerance, water and nitrogen use efficiency, and salt tolerance fall into this category. **Quality traits** are those that improve crop value to the consumer, whether the crops will be used to feed people or livestock (Figure 9–19). An example of quality trait improvement is increasing essential vitamins and nutrients, as in Golden Rice or BioCassava. Oil seed crops have also been improved to have healthier omega-3 fatty acids. The most recent traits of interest are those that allow use of crops as renewable energy resources or for biomanufacturing.

Courtesy of USDA Image #96c1794

Figure 9–19
Plants that produce feed for livestock can be genetically altered to increase the nutritional value.

Biofuels researchers must tackle the challenge of genetically engineering plants for use as feedstocks for microbial fermentation. Tough plant cell walls make for great fibers when weaving cloth or rope. However, the same cellulose microfibrils that make plant cell walls tough are difficult to break down. Cellulose microfibrils make up the largest available reservoir of sugar molecules on Earth (thanks to photosynthesis). The sugar molecules in cellulose are very difficult to break apart, but once available, they may be used by microbes as an energy source to ferment biofuels, such as ethanol. Ethanol made from crop residues or **biomass** is called **cellulosic ethanol**. Plants and algae may also be engineered to produce oils that may be used as **biodiesel** fuel.

In addition to making biofuels, researchers are researching plants that biomanufacture useful proteins and industrial precursors. Using genetically modified crop plants or plant cell cultures, it is possible to produce a number of useful proteins and oils, including **plant-made pharmaceuticals**. Human pro-insulin has been made in canola seed, tobacco cells have been used to produce vaccines and other medicines, and carrot cell cultures are used to make the complex protein treatments for Gaucher disease and Fabry disease. A wide range of pharmaceuticals can be biomanufactured by placing medicine-producing genes into plants. The medical benefits of plant biotechnology will be thoroughly discussed in Chapter 12. The broad spectrum of the current and soon-to-be-available biotech crops is summarized in the following table.

TABLE 9–1 Agricultural Biotech Products on the Market

Canola

LibertyLink Canola (Developed by Aventis CropScience)—Introduced in 1995, LibertyLink Canola allows growers to apply Liberty herbicide over the top during the growing season. This results in weed control with no effect on crop performance or yield.

InVigor Hybrid Canola (Developed by Aventis CropScience)—InVigor hybrid canola is made using a manageable and effective hybridization system that enables production of high-yielding hybrid canola varieties for sale to growers. InVigor hybrid seed was first sold in Canada in 1996 and in the United States in 2000.

Roundup Ready Canola (Developed by Monsanto)—Roundup Ready canola allows growers to apply Roundup herbicide over the top of the crop during the growing season for superior weed control with enhanced crop safety.

SMART Canola Seed (Developed by American Cyanamid seed partners)—Introduced in 1995, imidazolinone-tolerant canola allows growers to apply environmentally friendly imidazolinone herbicides to canola. In Canada, registration of ODYSSEY herbicide, a new imidazolinone for use on imidazolinone-tolerant canola, was approved on April 4, 1997. One postemergence application of ODYSSEY herbicide provides both contact and residual control of hard-to-control broadleaf and grassy weeds, resulting in maximum yield potential.

Corn

Attribute Bt Sweet Corn (Developed by Syngenta Seeds)—Attribute insect-protected sweet corn varieties from Syngenta provide a high level of built-in protection against European corn borer and corn earworm, protecting crops from ear damage and yield loss.

CLEARFIELD Corn (Developed by American Cyanamid seed partners, including Prozar, Myrogen, and Garst Boku 12)—Introduced in 1992, imidazolinone-tolerant and -resistant corn allows growers to apply the flexible and environmentally friendly imidazolinone herbicides to corn.

Registration of LIGHTNING herbicide, a new imidazolinone specifically for use on CLEARFIELD Corn, was approved by the EPA on March 31, 1997. One postemergence application of LIGHTNING herbicide provides both contact and residual control of broadleaf and grassy weeds, resulting in maximum yield potential.

(continued)

TABLE 9–1 Agricultural Biotech Products on the Market (*continued*)

LibertyLink Corn (Developed by Aventis CropScience)—Introduced in 1997 in the United States and 1998 in Canada, LibertyLink Corn allows growers to apply Liberty herbicide over the top during the growing season. Liberty herbicide kills over 100 grass and broadleaf weeds fast, with no crop injury. LibertyLink Corn hybrids are offered by seed company partners like Pioneer, Syngenta, Cargill, and Garst and over 100 other seed companies.

Liberty herbicide (Offered by Aventis CropScience)—NK Knockout Corn, NK YieldGard Hybrid Corn (Developed by Syngenta Seeds)—Syngenta Seeds has produced several corn varieties that have been modified to provide natural protection against certain pests.

Roundup Ready Corn (Developed by Monsanto)—Approved in 1997, Roundup Ready Corn allows over-the-top applications of Roundup herbicide during the growing season for superior weed control.

Cotton

Bollgard Insect-Protected Cotton (Developed by Monsanto)—Introduced in 1996, cotton with Monsanto's Bollgard gene is protected against cotton bollworms, pink bollworms, and tobacco budworms.

Roundup Ready Cotton (Developed by Monsanto)—Approved in 1996, Roundup Ready cotton tolerates both topical and post-directed applications of Roundup herbicide.

Peanuts

High Oleic Peanut (Developed by Mycogen)—Peanut plants modified by mutagenesis to produce nuts in high oleic acid results in longer life for nuts, candy, and peanut butter.

Potatoes

NewLeaf Insect-Protected Potato (Developed by Monsanto)—Introduced in 1995, the NewLeaf Potato is the first commercial crop to be protected against an insect pest through biotechnology. Thanks to a gene from a variety of the Bt bacterium, the NewLeaf Potato is resistant to the Colorado potato beetle.

NewLeaf Plus (Developed by Monsanto)—Insect- and virus-protected potatoes. These potatoes are protecting themselves against Colorado potato beetles and potato leaf roll virus.

New-Leaf Y Insect- and Virus-Protected Potatoes (Developed by Monsanto)—These potatoes protect themselves against the Colorado potato beetle and the potato virus Y.

(continued)

TABLE 9–1 Agricultural Biotech Products on the Market (*continued*)

Rapeseed

Laurical (Developed by Calgene, LLC)—A less expensive source of high-quality raw materials for soaps, detergents, and cocoa butter replacement fats. Rapeseed plants with more than 45 percent laurate in oil have been produced.

Soybeans

Novartis Seeds Roundup Ready Soybeans (Developed by Novartis Seeds); Roundup Ready Soybeans (Developed by Monsanto)—Introduced in 1996, Roundup Ready Soybeans allow growers to apply Roundup herbicide over the top during growing season. The result is dependable, superior weed control with no effect on crop performance or yield.

Sunflowers

High Oleic Sunflower (Developed by Mycogen)—Sunflower plants modified by mutagenesis to produce sunflower oil that is low in trans-fatty acids, does not require hydrogenation, and has improved temperature stability.

Tomatoes

Increased Pectin Tomatoes (Developed by Zeneca Plant Sciences)—Tomatoes that have been genetically modified to remain firm longer and retain pectin during processing into tomato paste.

Miscellaneous

Messenger (Developed by EDEN Bioscience)—This is the first of a series of products based on naturally occurring harp in protein technology. Approved by the EPA in April 2000, Messenger stimulates growth and defense pathways inherent within each plant without altering the plant's DNA. Messenger treatments promote healthier plants and increased yields as well as increased disease resistance and deterrence of insects such as nematodes. Messenger is a labeled product, currently being sold in cotton, citrus, apples, strawberries, rice, tomatoes, peppers, cucurbit vegetables, cane berries, grass seed, potatoes, and many other crops.

ON THE MARKET WITHIN 6 YEARS

Alfalfa

Roundup Ready Alfalfa (Developed by Monsanto)—Allows over-the-top applications of Roundup herbicide during the growing season for superior weed control.

Apples

Bt Insect-Protected Apple (Developed by Monsanto)—These apples will contain built-in insect protection against codling moth.

(continued)

TABLE 9–1 Agricultural Biotech Products on the Market (*continued*)

Bananas

Disease-Resistant Bananas (Developed by DNA Plant Technology Corporation)—These bananas will be resistant to the fungal disease black sigatoka.

Corn

Insect-Resistant Corn (Developed by Dow AgroSciences and Pioneer Hi-Bred International, Inc.)—These corn hybrids will provide broader-spectrum control of insect pests than what is currently available, including first- and second-generation European corn borer, southwestern corn borer, black cutworm, and fall armyworm.

Rootworm-Protected Corn (Developed by Monsanto)—This corn carries built-in protection against corn rootworm. In addition, future products may be stacked with Yieldgard, Roundup Ready, or both to provide broader pest protection.

Rootworm-Resistant Corn (Developed by Dow AgroSciences & Pioneer)—These new hybrids will produce a protein that is toxic to the corn rootworm, thus eliminating or reducing the need for soil-applied insecticides.

Second-Generation Yieldgard Insect-Protected Corn (Developed by Monsanto)—This corn is protected against insect pests like the original Yieldgard corn but using a different mode of action to help growers cover a broader spectrum of insect pests and better manage insect resistance concerns.

Cotton

Insect-Protected Cotton (Developed by Dow AgroSciences)—This cotton will provide insect resistance that is broader in spectrum than what is currently available.

Second-Generation Bollgard Insect-Protected Cotton (Developed by Monsanto)—This cotton controls insect pests, like the original Bollgard cotton, but using a different mode of action to help growers manage insect resistance concerns.

Lettuce

Roundup Ready Lettuce (Developed by Monsanto)—Allows over-the-top applications of Roundup herbicide during the growing season for superior weed control.

Potatoes

Bruise-Free Potatoes (Developed by Monsanto)—These potatoes will not form internal bruises or black spots but will keep their bright color after peeling and be much more appealing to the consumer.

(continued)

TABLE 9–1 Agricultural Biotech Products on the Market (*continued*)

High-Solids Potato (Developed by Monsanto)—Monsanto has developed a higher-solids (or starch content) potato by introducing a starch-producing gene from a soil bacterium into a potato plant. With the reduction in the percentage of water in the genetically improved potato, less oil is absorbed during processing, resulting in a reduction of cooking time and costs, better-tasting french fries, and an economic benefit to the processor.

NewLeaf Multi-Virus Resistant, Insect-Protected, Roundup Ready Potatoes (Developed by Monsanto)—These potatoes will combine the benefits of resistance to potato leaf roll virus, potato virus Y, and Colorado potato beetle with the benefits of Roundup Ready weed control.

Roundup Ready Potatoes (Developed by Monsanto)—These potatoes will allow growers to apply Roundup herbicide during the growing season for improved weed control and crop safety.

Rice

CLEARFIELD Rice Seed (Developed by American Cyanamid)—American Cyanamid is cooperating with universities and public and private seed companies to develop rice varieties tolerant to imidazolinone herbicides. Imidazolinone herbicides are flexible, are environmentally friendly, and provide superior contact and residual control of weeds.

Soybeans

Insect-Protected Soybeans (Developed by Monsanto)—These soybeans will contain built-in insect protection from key soybean insect pests.

Strawberries

Strawberry (Developed by DNA Plant Technology Corporation)—The company is adding genes to confer resistance to glyphosate herbicide and fungal diseases.

Sugar Beets

CLEARFIELD Sugar Beet Seed (Developed by American Cyanamid)—American Cyanamid is cooperating with universities and seed companies to develop sugar beet varieties tolerant to imidazolinone herbicides. Imidazolinone herbicides are flexible, are environmentally friendly, and provide superior contact and residual control of weeds.

Roundup Ready Sugar Beets (Developed by Monsanto)—Roundup Ready sugar beets are tolerant of Roundup herbicide and provide growers with a new weed control option while the crop is growing.

Tomatoes

Roundup Ready Tomato (Developed by Monsanto)—Allows over-the-top applications of Roundup herbicide during the growing season for superior weed control.

(*continued*)

TABLE 9–1 Agricultural Biotech Products on the Market (*continued*)

Wheat

CLEARFIELD Wheat Seed (Developed by American Cyanamid)—American Cyanamid is cooperating with universities, public and private laboratories, and seed companies to develop wheat varieties tolerant to imidazolinone herbicides. Imidazolinone herbicides are flexible, are environmentally friendly, and provide contact and residual control of weeds common to wheat production, including ones not controlled by currently registered wheat herbicides.

Miscellaneous

Genetically Modified Fruits and Vegetables with Longer Postharvest Shelf Life (Developed by Agritope, Inc., a wholly owned subsidiary of Epitope, Inc.)—Using ethylene control technology, Agritope, Inc. has created delayed-ripening, longer-lasting tomatoes and raspberries.

LibertyLink Soybean, Cotton, Sugar Beet, and Rice (Developed by Aventis Crop-Science)—These LibertyLink crops will be available in Canada and/or the United States. Like LibertyLink Corn, when used together with Liberty herbicide, they will allow farmers greater flexibility and environmental soundness in weed control.

Courtesy of the Council of Agricultural Education.

Summary

Humans have genetically modified plants for thousands of years. Until recently, these changes have come about as a result of selective breeding. The advent of genetic engineering has greatly accelerated the process of plant improvement. Now scientists can select the specific desirable genes, insert them into conventional crops, and produce crops that have traits not easily obtained by conventional plant breeding. A variety of traits may be genetically engineered in crops plants. Agronomic trait engineering to meet the needs of an expanding global population will help us produce more food on the same amount of arable land, keeping in mind sustainable environmental practices. Quality trait engineering will help to alleviate nutritional deficiencies in populations with limited food selections. Renewable resource traits will allow us to use nonarable land to produce biofuels. Plants used for biomanufacturing make production of pharmaceuticals and industrial precursors more efficient and less expensive, with benefits passed on to consumers. Ongoing research in plant biotechnology and agriculture will help us to tackle challenges in health care, energy, and the preservation of natural resources.

CHAPTER REVIEW

Student Learning Activities

1. Find out if there are any genetically engineered crops growing in your area. Your county Cooperative Extension Service office would be a good place to find the information. If possible, interview the producers of the crops and find out their opinions of genetically modified crops. Share your findings with the class.

2. Conduct a Web search on genetically modified crops. Find materials written by people who oppose the growing of these crops. Outline your opinion of the articles. Do you feel that the concerns are founded? Why or why not?

3. What do you consider to be the most important genetically modified plant in existence? Give reasons to support your idea.

Fill in the Blanks

1. Even with the advent of today's highly effective _____ and management techniques, it is estimated that about _____ percent of the world's food supply is lost to _____, _____, and _____.

2. Scientists have known for many years that the bacteria _____ _____ (commonly known as _____) was effective in _____ insects by disrupting their _____ processes.

3. _____ who have an opportunity to try _____ crops often see a great _____ benefit and adopt this new technology, but other members of the _____ are slower to appreciate the _____.

4. Another serious problem faced by producers are _____ that _____ with crops for soil _____ and _____.

5. In plants that have been engineered to resist _____ infections, a gene for the viral _____ coat is place in the plant _____, a strategy referred to as _____ _____ resistance.

6. During the _____, the _____ _____ included the development of crop varieties that could be grown in specific _____ regions.

7. By studying the biology of _____ _____ plants and those that are "salt loving" or _____, researchers have identified genes that help salt _____ crops survive highly _____ soil conditions.

8. _____ that increase crop _____ or make crops less _____ to grow are called _____ traits.

9. Ethanol made from crop _____ or _____ is called _____ _____.

10. A wide range of _____ can be _____ by placing medicine-producing genes into _____.

True or False

1. Interrupting an insect's reproductive cycle is an effective way to control pests using biotechnology.

2. Bt toxins are effective at killing birds and mammals.

3. Most applied pesticides are effective for long periods of time.

4. The insect that has caused the most damage to corn crops is the boll weevil.

5. Roundup is a nonselective herbicide.

6. Stress tolerance genes can be used to improve plant growth and crop yields.

7. Plants and algae can be engineered to produce oils that can be used as fossil fuels.

8. Plants with nitrogen use efficiency traits require less chemical fertilizers than other plants.

9. Cellulose is a good source of sugar because cellulose molecules are very easy to break apart.

10. Thousands of acres of arable land are lost each year to increased soil salinity levels.

Discussion

1. Explain the advantages of genetically engineering crops that are resistant to insect pests.

2. List the crops that have been produced using the Bt gene.

3. Explain why planting crops that are genetically engineered using the Bt gene is more efficient than using chemical pesticides.

4. Describe the drawbacks of using systemic pesticides.

5. Explain how using the Bt gene can help reduce the amount of mycotoxins found in corn crops.

6. Distinguish between selective and nonselective herbicides.

7. What does it mean to say that a crop has "stacked traits"?

8. Explain what is meant by "quality traits"? Give two examples of crops with quality traits.

9. How are plants used to biomanufacture pharmaceuticals?

10. Select three agricultural biotech products that are currently on the market and explain the features of each product.

CHAPTER 10

Biotechnology in Animal Reproduction

KEY TERMS

artificial insemination
sperm
semen
sire
quarantine
protectant
estrus
artificial vagina
ejaculation
conception
motility
extenders
straws
hormones
estrous cycle
follicle-stimulating
  hormone (FSH)
follicle
embryo transfer
dam
progeny testing
donor cows
recipient cows
superovulation
gonadotropin releasing
  hormone (GnRH)
luteinizing hormone (LH)
corpus luteum
sperm sexing
flow cytometer

OBJECTIVES

When you have finished studying this chapter, you should be able to:

- Explain how biotechnology is used in animal breeding.
- Describe the advantages of artificial insemination.
- Explain the process of artificial insemination.
- Explain how semen is processed for storage and shipment.
- Explain the process of estrus synchronization.
- Describe how eggs are produced in the ovary.
- Discuss the advantages of embryo transfer.
- Tell how embryo transfer is accomplished.
- Relate the advantages of producing only male or only female offspring.
- Explain how sperm are sorted according to sex chromosomes.

Introduction

Humans began breeding animals as soon as animals were domesticated, and they understood that it took both a male and a female to mate to produce an offspring. Although people began to select those males and females with the most desirable characteristics to use for breeding animals, until relatively recently they had no idea how the process of conception took place. Over the years by acquiring knowledge through research, humans began to discover how animal reproduction occurs. As this new knowledge was generated, it was applied to make the breeding process more efficient in producing the type of animals desired.

Every aspect of the livestock industry is directly dependent on the reproduction of animals (Figure 10–1). We have come to depend on many products from animals for a wide variety of uses in our everyday lives. Without the efficient reproduction of agricultural animals, this supply would cease. All the processes used in improving the efficiency of reproduction involve the use of biotechnology. In another chapter, the use of cloning in the reproduction process was discussed. In this chapter, the more conventional uses of biotechnology in animal breeding are discussed.

Figure 10–1
The reproduction of animals is of central importance to the livestock industry.

Artificial Insemination

One of the most used aspects of animal biotechnology is that of **artificial insemination** (AI). This process involves the introduction of the male **sperm** into the reproductive tract of a female animal by means other than the natural mating process. This technology is widely used in most segments of the livestock industry, with several large companies supplying **semen** for producer use. A high percentage of the superior breeding animals result from artificial insemination. Most of the dairy cows, most swine, and essentially all turkeys rely on the use of artificial insemination for impregnating females (Figure 10–2).

There are several distinct advantages of artificial insemination over that of allowing animals to breed naturally. The following outlines the advantages:

- When animals are bred, producers are always interested in using the highest-quality animal available. Almost always, high-quality **sires** (the father animals) from any of the species are expensive, and the animals may not be available at any cost. The cost of semen from artificial insemination is often much lower than that of buying a sire of the desired quality.
- For the sires having the most desirable traits, meticulous records are kept on their offspring. The data from these records, called progeny data, are used by the producer in determining the quality of the sire he or she is contemplating using. Since one bull may produce many

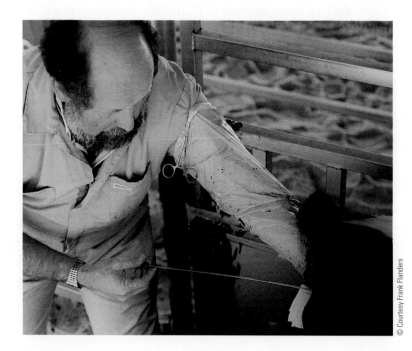

© Courtesy Frank Flanders

Figure 10–2
Most of the cows in the dairy industry are produced through artificial insemination.

thousands of offspring, progeny data can be a reliable way of predicting how the offspring of the sire will perform. The American Breeders Service reports that one of their superior bulls lived for 12 years and produced over 462,000 ampules of semen. The data from this bull were quite extensive and very useful in selecting a sire with certain characteristics.

• Producers have differing goals as to what they want to achieve with their livestock operations. Once the goals for the operation have been set, a clear-cut idea of the type of males and females is needed to produce the offspring desired. Artificial insemination allows the producer to select the type of sire needed for a particular group of females. For instance, a hog producer may need to increase the frame size of the animals in his or her herd, or a beef producer may need a bull that will sire smaller calves at birth for calving ease. Through the use of sire data, the producer can select sires that are known for these characteristics.

• By nature, most male animals are aggressive. This characteristic has been developed by nature to ensure that the strongest, healthiest, and most virile male is the one to breed the females. This can cause problems for producers who keep male animals for breeding because the animals can be dangerous (Figure 10–3). Since the use of artificial insemination (AI) became common, the U.S. Department of Agriculture no longer maintains statistics

Figure 10–3
Male animals can be aggressive and dangerous.

on the number of farmers killed annually by bulls as it once did. Through the use of artificial insemination, producers do not have to keep male animals.

- A real problem that faces livestock producers is that of animal diseases. Diseases can be transmitted in several ways, and many diseases can be transmitted through the mating process. Brucellosis, leptospirosis, trichomoniasis, and vibriosis are all examples of serious diseases that can be spread by sexual contact. The proper use of artificial insemination almost totally eliminates the likelihood of disease being transmitted through mating.

- Livestock is produced all over the world, and a very high-quality sire may be produced in another country. If animals are imported, they must go through a **quarantine** process that requires the animals to be kept in isolation for a period of time to make sure that they do not bring disease into the country. This process is not only very expensive, but can take several months. By using frozen semen, new genetics can be brought into the United States with less risk of importing disease and at much less cost.

- If a producer owns his or her own sires, the expense of replacing sires is substantial. In addition, if the producer is not pleased with the offspring of the sire, then the old sire has to be sold and a new one bought. By using artificial insemination, the producer only has to order semen from a different sire to change sires. Furthermore, replacement females can be raised and a different sire selected for them.

The Development of Artificial Insemination

Historians say that artificial insemination has been used for hundreds of years. In fact, some say that the process goes all the way back to the Middle Ages. Legend is that warring Arab tribes slipped into enemy camps and collected semen from stallions that belonged to their enemies. The raiding tribesmen used the stolen semen to breed their mares and produce superior foals (Figure 10–4). We do not have very much documentation that this really took place since the stories were told orally and were passed down from generation to generation. However, it is entirely possible that artificial insemination began in this way.

Figure 10–4
Legend says that the first artificial insemination was used by the Arabs during the Middle Ages.

Figure 10–5
During World War I, horses were in great demand. To help meet the need, a Russian scientist used artificial insemination to rapidly produce horses.

Documentation of the first successful use of artificial insemination was in 1780 by an Italian scientist named Lazarro Spallanzani, who was successful in artificially inseminating dogs. This procedure saw limited use and had very little economic value. Perhaps the first large-scale use of artificial insemination was by the Russians shortly after the turn of the twentieth century. During World War I, horses were still used for pulling wagons and artillery and for carrying soldiers in the cavalry. During the awful years of fighting, most of the horse population of Europe was destroyed, and horses were very much in demand. To help meet the demand for thousands of horses, a Russian physiologist named Ivanoff used artificial insemination as a means of rapidly producing foals from superior sires (Figure 10–5). This same technology used on cattle and sheep became an accepted practice in the 1920s and 1930s.

The biggest problem in using artificial insemination was the preservation of the live sperm. The life of sperm is only about 2–3 days, so there were problems in obtaining semen when it was needed. The idea of freezing semen to preserve it was developed when the use of artificial insemination had been perfected. However, the problem was that a relatively small number of the live sperm survived the freezing process.

After several years of research, a successful technique was developed in the 1950s. The newly developed technology involved the use a **protectant**, such as glycerine, that is added before the semen is frozen (Figure 10–6). The temperature of the mixture is lowered at a specific steady rate until the temperature reaches −320°F. Semen kept at this temperature can remain viable for years. In fact, bull semen has been successfully stored for as long as 30 years. Semen from bulls, stallions, and rams can be frozen, stored, and thawed successfully. In the past, almost all semen from boars was shipped and used fresh because of problems with sperm livability when the semen is frozen. However, new techniques now allow the use of frozen semen. While fresh semen usually gives a higher conception rate, the use of frozen semen is now commercially feasible.

One of the drawbacks of using artificial insemination is that it is more labor intensive than using natural mating. Females have to be closely monitored for signs of **estrus** (the period when she is ready for mating), and they also have to be restrained and inseminated one at a time (Figure 10–7). This is why artificial insemination is used most widely in the dairy industry. Dairy cattle have to be brought into a milking parlor twice a day, and it is easy to monitor and restrain them. On the other hand, beef cows usually roam in a large pasture and are not penned on a daily basis. This makes them more difficult to breed by artificial insemination.

Semen Collection and Processing

The starting point of artificial insemination technology is collecting the semen from healthy male animals. Healthy males must be maintained to ensure a high-quality collection

Figure 10–6
A protectant is added to the semen before it is frozen.

Figure 10–7
Females have to be closely monitored for signs of estrus. This demands a lot of time.

that is free from disease-causing organisms. This process makes use of a thorough understanding of animal physiology, anatomy, and behavior. Semen is collected through the use of a dummy animal or another live animal and an artificial vagina. The **artificial vagina** consists of a rigid tube that is lined with a smooth-surface water jacket that is filled with warm water (Figure 10–8). At the end, a test tube is attached for collecting the semen. As the male approaches and mounts the dummy or live animal, the penis is guided into the artificial vagina, where **ejaculation** occurs. The amount of the ejaculate or semen varies with the different species.

Technicians must make certain that the semen is of sufficient quality to survive freezing and thawing with enough viable sperm to complete **conception**. Once the semen is collected, it is examined in the laboratory under a microscope for foreign material and for quality (Figure 10–9). Quality is determined by the number of sperm in a milliliter of semen, how active the sperm are (**motility**), and the shape of the sperm. Sperm motility is a measure of how active sperm are. Very active sperm are desirable because of the distance they must travel to reach the oviduct of the female. In a quality collection, a high percentage of sperm should be motile, that is, moving purposefully in a given direction rather than simply moving around in circles. Immotile sperm will not be able to fertilize the egg and may interfere with the healthy, motile sperm. A high percentage of immotile

Figure 10–8
Semen is collected using an artificial vagina.

© Courtesy of Michigan State University

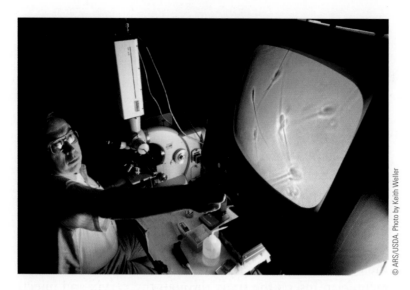

Figure 10–9
Sperm is evaluated for number, motility, and shape. This technician is using a microscope and a video monitor to do the evaluation.

sperm is usually an indication of a problem with either the physiology of the male or the collection procedure. Sperm of different species are shaped differently. The sperm are checked to make sure the shape is normal. A large number of sperm with an unusual shape is not desirable.

After a thorough examination, the sperm are processed. Processing involves adding **extenders**, such as milk, egg yolk, glycerine, and/or antibiotics. Since one ejaculation can contain billions of viable sperm, one purpose of the extenders is to provide a means of diluting the semen (Figure 10-10). The semen from one bull ejaculation may be divided into several units, depending on the number of sperm in the ejaculate. By doing this, several females can be bred from one ejaculation. Another purpose for adding extenders is to provide protection to the sperm during the freezing procedure. Extenders also provide nourishment for the sperm.

The semen is checked again after the addition of the extenders to make sure that the sperm are still motile. The semen is then packaged in small hollow tubes called **straws**, sealed, and labeled with the name of the company, the date, and the name of the sire. The straws containing the semen are frozen at a specific rate to approximately −320°F and stored and transported in liquid nitrogen tanks (Figure 10–11).

The semen is thawed immediately before it is placed into the female tract. When the technician is ready to artificially inseminate an animal, the straws are carefully removed from the liquid nitrogen tank. Precautions have to be taken because the liquid nitrogen can cause a frost bite–like

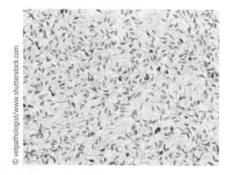

Figure 10–10
Since one ejaculation can contain billions of live sperm, the semen can be diluted. Note that the slide has been stained to aid visibility.

Figure 10–11
Semen is packaged in straws and frozen.

injury if it contacts the skin. The straws of semen have to be thawed at the proper temperature and speed. Thawing may be accomplished through the use of a special apparatus that heats water to a certain temperature or through the use of a water-filled thermos bottle. The straw is placed into the water and left for not less than 30 s and not more than 15 min. Proper thawing ensures that the thawed sperm will be healthy and motile.

Once the semen is properly thawed, the straw containing the semen is placed in a tube-like instrument that will be used to place the semen in the tract of the female. This is accomplished by a technician extending his or her hand into the rectum and colon to locate the cervix. The cervix is located by grasping the female tract through the wall of the colon. Once the cervix is located and held in place, the technician inserts the straw through the cervix and injects the sperm (Figure 10–12). After the semen is placed in the female tract, the process of fertilization takes place just as in natural mating. The person doing the insemination must undergo special training before he or she can develop the skills necessary to properly thaw the semen and place it correctly.

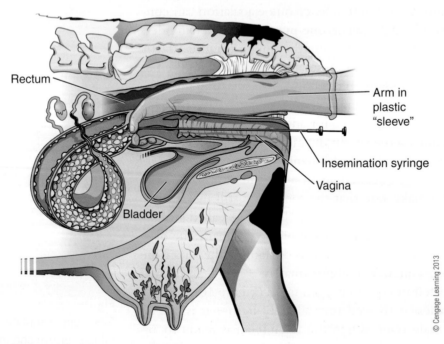

Figure 10–12
The thawed semen is placed using a tube that is inserted through the cervix.

Control of the Estrous Cycle

As mentioned previously, a disadvantage of artificial insemination is that females have to be monitored for signs of estrus. Obviously, if a whole herd of females come into estrus at the same time, labor and time costs can be greatly decreased. A biotechnology advance that has greatly aided in improving the reproductive efficiency of agricultural animals is the use of estrus or heat control. Estrus is the time when the female allows breeding, and it is controlled by the production and secretion of **hormones** at the proper time. The **estrous cycle** is a chain of events that occur as certain hormones are released. If hormones from an outside source are introduced into the female, the hormone will cause the same reaction as a naturally produced hormone. For example, **follicle-stimulating hormone (FSH)** causes the **follicle** to develop on the ovary, and the egg is developed within the follicle. Injecting females with the proper hormones will cause the female to begin the cycle at that point and come into estrus on schedule. By inducing estrus using hormones on all the females in the herd at the same time, the cycle can be synchronized so that all the animals come into heat at about the same time. These hormones are provided by a variety of methods, including injections, but more commonly through feed additives and uterine implants. The obvious advantage is that a producer can have all the animals in a herd artificially inseminated at the same time. This saves time and resources not only at breeding but also when the females calve or farrow at about the same time (Figure 10–13). The crop of young animals will be of about the same age, so they can be managed in the same way as they are grown out.

Embryo Transfer

Artificial insemination allows the widespread use of genetics from superior males. As the improved genetics become apparent in the offspring, only approximately half the superior genetics show up in the males. It is easier to multiply the effects of the male since the male naturally produces billions of sperm on a regular basis. This makes males capable of producing thousands of offspring each year. However, most females, such as cows, may produce only one offspring per year. Even

Animal Gene Collections Support United States Research

The National Animal Germplasm Program (NAGP) opened in Fort Collins, Colorado, in 2000, with genetic material from 40 chicken lines. Since then, the collection has expanded to include dairy and beef cattle, swine, sheep, goats, bison, elk, and fish. Today, NAGP houses more than 547,000 samples of genetic material, or germplasm, from more than 12,000 animals. This collection—like all Agricultural Research Service collections of animal germplasm—preserves the genetic diversity of agriculturally important animals.

Providing genetic material for genomic studies is one of the most important functions these collections serve. NAGP, for example, has distributed samples from about 2,500 animals to ARS researchers and their university colleagues. ARS scientists have used bull semen acquired from NAGP to genotype prominent bulls that had sired dairy cattle. This information, combined with milk-production data gathered from those cows, has been used to improve dairy cattle breeding programs. Similar work has been done with beef cattle and pigs by ARS researchers in Clay Center, Nebraska.

In many cases, these collections have helped united States animal producers save money, but ARS animal collections are not simply economical. They provide information for researchers and breeding material for animal producers, and they play an instrumental role in protecting and improving agricultural livestock.

Jurassic Pork: Protecting and Promoting Rare Agricultural Breeds

Animal genetic material cannot raise the dead. But it can be used to revive animal lines that have died out.

© USDA/ARS #K10187-9. Photo by Stephen Ausmus

National Animal Germplasm Program coordinator Harvey Blackburn and technician Ginny Schmit place germplasm samples into a liquid nitrogen tank for long-term storage.

Researchers at Purdue University arranged to save semen from a unique line of pigs that had two mutations that negatively influenced meat quality. Maintaining live pigs is expensive and time-consuming, so the population was terminated when the studies involving the herd concluded. At the request of Purdue researcher Terry Stewart, NAGP scientists gathered semen from three of the boars in the study and added the samples to their extensive collection of animal genetic materials.

A few years later, Stewart and his colleagues decided to take the research in a new direction. By that point, the original herd was long gone, but the scientists were able to resurrect the line by inseminating seven sows with the saved semen. All the sows became pregnant and bore litters, indicating that in some instances, cryopreserving genetic material may be a more efficient use of time and money than maintaining a live herd.

"This is the first time a line has been cryopreserved, discontinued, and reestablished using germplasm frozen and stored by NAGP," says animal geneticist and NAGP coordinator Harvey Blackburn.

The NAGP collection has also been used to prevent existing lines from dying out. For example, scientists at NAGP have collected and distributed germplasm to support a rare breed of dairy cattle.

Shorthorn cattle first came to the United States in the 19th century from the United Kingdom. Imports increased significantly after the breed was successfully crossed with the Texas Longhorn. The exact number of milking and beef Shorthorns in the United States today is unknown, but in 2000, there were about 2,800 milking Shorthorns registered. Though they have lower milk yields than Holstein or Jersey cows, Shorthorns are valued for their calm temperament and production efficiency.

Shorthorns have been successfully crossbred with many other dairy and beef cattle breeds. These crosses have performed well, but the breed's fans lobbied for the establishment of a program to identify purebred native Shorthorns.

NAGP scientists have made special efforts to collect genetic material from cattle that could be designated as native Shorthorns. In 2006, NAGP animal physiologist Phil Purdy traveled to Nebraska to collect genetic material from 15 bulls in a large Shorthorn herd.

"This herd is of particular importance because, except for one animal, no outside animals have been brought in for breeding since before World War II," Blackburn says. In addition, Shorthorn samples from the repository have been used by breeders in Utah to introduce new genetic variability into the Shorthorn breed.

—This research is part of Animal Health, an ARS national program (#103) described at http://www.nps.ars.usda.gov.

Figure 10–13
If cows can be bred at the same time, the calves will be born about the same time. This is a great benefit to the producer.

Figure 10–14
Embryo transfer allows for greater use of genetics from superior females.

pigs usually have only about two litters per year. Scientists tried for years to develop the biotechnology necessary to expand the genetics of superior females. Today, the transferring of embryos from one female to another is common (Figure 10–14). A female is capable of producing many thousands of eggs during her lifetime, although only a relative few develop into offspring. If the eggs are collected from a superior female and implanted into an inferior animal, the superior female has the capacity to produce many offspring in a year.

The biotechnology of **embryo transfer** has many benefits:

- As already mentioned, the use of embryo transfer allows the rapid advancement of genetics from the **dam** (the mother animal). Just as artificial insemination allows the production of many offspring from a superior male, embryo transfer allows the production of many offspring from a superior female.
- Embryo transfer allows the **progeny testing** of females. Progeny testing involves the gathering of data from the offspring of a particular animal. The data are analyzed to determine how valuable the animal is as a parent. Through the use of artificial insemination, a male can be progeny tested in a short time because of the tremendous number of offspring that can be born and raised at the same time. The problem with progeny testing females

© Cindy Flower. Image from BigStockPhoto.com.

Figure 10–15
Through the use of embryo transfer, one female can produce several calves each year.

is that during the lifetime of a female, her offspring are very limited in number, and this does not allow sufficient numbers of offspring from which data can be collected. Through the use of embryo transfer, one female can produce many offspring in a short period of time, allowing for the testing of her progeny (Figure 10–15).

- As in artificial insemination, embryo transfer permits the import and export of quality animals without the quarantine measures required of animals that are already born.

- Embryo transfer allows the use of a dual production system. For example, the reason that dairy cows are bred is so that they will continue to produce milk. If they are bred naturally or using artificial insemination, the calves will still be half dairy animals. By using embryo transfer, dairy cattle can produce calves that are pure beef animals.

- By implanting two embryos into a recipient female, twin offspring can be produced, though this may not be advisable in animals that do not typically bear twins (cows and mares).

- A producer can rapidly convert his or her herd from grade animals to a purebred herd. By implanting the female of mixed breeding with purebred embryos, a producer can raise replacement animals that are both purebred and of high quality.

Some argue that the use of embryo transfer has a big disadvantage. They say that as producers use embryo transfer and artificial insemination over a period of many years, the genetic base of the various breeds of animals will narrow. This means that, in time, there will be only a relative few animals that will eventually provide genetic material (egg and sperm) for the perpetuation of the breed. The fear is that producers, by demanding only embryos from the best animals, will eventually cause the loss of animals that the producers feel are inferior. If the only existing animals of a breed are related, there is reason to believe that this will lead to the weakening rather than the strengthening of the breed. Others contend that through the use of embryo transfer, the importation of genetic strains from all over the world will prevent this problem from occurring. They contend that there are enough different strains in the world to make a narrowing of the genetic base highly improbable.

The Process of Embryo Transfer

The process of embryo transfer begins with the selection of **donor cows** and **recipient cows**. Cows selected as donors are usually animals that are of unusual value as breeding animals. They possess characteristics that are highly desirable to pass on to offspring. These characteristics might include high milking ability, ability to grow, or reproductive capacity. Or the animals might be the type in demand for the show ring. In any case, the donor animals are too valuable to produce an offspring only once a year. Producers may purchase frozen embryos from one of many companies that specialize in the sale of genetically superior embryos. The producer selects the embryos he or she wishes to order by analyzing data that have been compiled about the donor and the sire. These data usually consist of production data of the animal's ancestors and their progeny. In this way, the producer can select those traits that will be of most use in the producer's herd. On the other hand, recipient cows (those into which the embryo will be transferred) are usually cows of ordinary value, although they are also carefully selected (Figure 10–16). They must be healthy animals that are able to reproduce efficiently. That is, they must have the ability to maintain pregnancy and deliver a healthy, growing calf at the end of the gestation period.

© Courtesy of ARS/USDA. Photo by Scott Bauer

Figure 10–16
Recipient cows must also be carefully selected. This is a Senopol surrogate mother with a Romosinuano embryo transfer calf.

After the donor and recipient animals are selected, both groups must be synchronized so that they are at the same phase of their estrous cycle. This allows for the proper transfer of the embryo from one reproductive system to another. This synchronization is accomplished using the procedures discussed earlier. The only difference is that the donor animals undergo a process known as **superovulation**, causing the donor animal to release several eggs instead of just one. In this way, as many as 12–15 eggs or more can be collected from one ovulation. Superovulation is accomplished by injecting the donor with FSH. This hormone causes the ovaries to produce several follicles instead of one. Eggs develop and mature within these follicles. Two to three days later, the female is injected with **gonadotropin releasing hormone** (GnRH) or **luteinizing hormone** (LH) to cause her to come into estrus. About 48 h later, the female should be in estrus, or heat. When this occurs, the cows are artificially inseminated or bred naturally.

Once fertilization occurs, the fertilized eggs (embryos) are allowed to grow for about a week and are then collected. In the earlier days of embryo transfer, collection was accomplished by removing the embryos surgically. This caused problems because of the scarring of the tissue in the reproductive tract of the donor female. Today, the embryos are removed through a process called flushing. In this procedure a long, thin rubber tube called a catheter is passed through the cervix and into the uterine horn (Figure 10–17). The catheter has an inflatable bulb

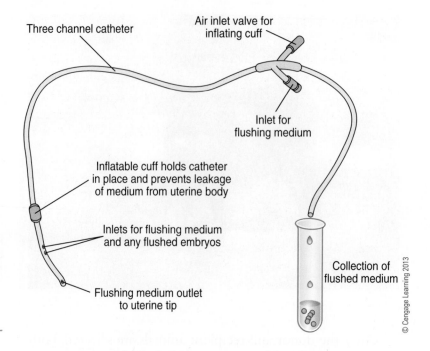

Three channel catheter

Air inlet valve for inflating cuff

Inlet for flushing medium

Inflatable cuff holds catheter in place and prevents leakage of medium from uterine body

Inlets for flushing medium and any flushed embryos

Flushing medium outlet to uterine tip

Collection of flushed medium

© Cengage Learning 2013

Figure 10–17
After a female has superovulated, a catheter is used to collect the embryos.

about two inches from the end that fills like a balloon and seals the entrance to the uterus. A solution is then injected through the catheter into the fallopian tubes. When the fallopian tubes and the uterus are filled with solution, the flow of the solution is stopped, and the solution is drained off into a collection cylinder. The fertilized eggs (embryos) are carried out of the uterus with the solution. An average of about six embryos is collected with each flush, though some animals will produce many more. After the embryos are flushed out, an injection of prostaglandin is administered to terminate any pregnancy that may result from any embryos that were missed. This process may be repeated at 3-month intervals (in the cow) if desired.

Once the embryos are collected in the solution, they are strained from the solution and examined under a microscope to determine their quality. Only embryos that are in the proper stage of maturity and that appear normal and undamaged are used for transferring. The embryos may be transferred directly to a recipient female or may be frozen and stored for implantation at a later date.

The recipient cow is made to come into estrus using a feed additive containing progesterone. When the **corpus luteum** reaches the proper stage, the embryo is placed in the uterus of the recipient cow. The transfer process must be properly and carefully performed to ensure cleanliness, as the uterus is very susceptible to infection at this stage, while it is much less susceptible at the earlier stage, when breeding would normally occur. The pregnancy is allowed to progress as it would in a normal conception. Research has shown that pregnancies from embryo transfer are as likely to go full term and result in delivery of a normal calf as are pregnancies from natural conception.

Sperm Sexing

A technology that promises to greatly benefit livestock producers allows them to produce only male or only female offspring. This is particularly important in certain aspects of the industry. For example, in the dairy industry, many more females are needed than males. Since only the females produce milk, very few males are wanted. In conventional breeding, about half the calves born are male, and these are often marketed as veal calves. Because dairy animals have been developed for milk production instead of meat production, male dairy calves are not considered to be efficient for producing beef. Breeders would greatly prefer to produce almost all female calves.

In the swine industry, pigs are produced for pork, and consumers demand high quality in pork products. Pork from boars has a strong odor and is usually considered to be tainted and unacceptable by consumers. To prevent this problem, young male pigs are castrated before they begin to mature sexually. This management practice is very effective; however, the procedure takes time, and the pigs lose growth during the healing process. These problems cost money for the producer but expenses could be saved if only gilt (female) pigs were produced. Not only would costs associated with castration be saved, but a group of pigs could be raised more efficiently. This is because gilts generally grow slightly faster and have leaner carcasses than barrows (castrated males).

Beef breeders who specialize in raising bulls for the purebred market could benefit if almost all the calves produced were males. In the purebred industry, bulls usually sell for more money than heifers, so an operation that produced almost all males might be more profitable. Those males lacking the quality for breeding animals could be raised for the beef market.

A problem in the poultry industry is that hatcheries producing chicks that will mature into layers may have half their chicks that are male. Since these chicks have been bred as layers,

Figure 10–18

Rooster chicks are of little value in layer operations.

Sex Determination in Animals

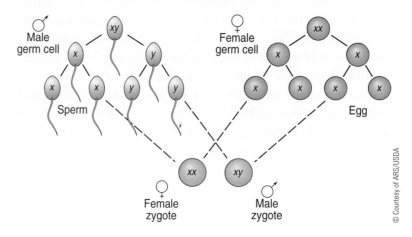

Figure 10–19

Male sex chromosomes have a Y shape, and female sex chromosomes have an X shape.

they lack the muscling to be used as broilers. Rooster (male) chicks are of little value to layer operations (Figure 10–18). If hatcheries could produce batches of chicks that are mostly female, a tremendous boost could be added to the efficiency of the industry. Conventionally produced chicks must be sexed in a procedure that is expensive and time consuming.

Remember from studying about genetics that sex in mammals is determined by the matching of the sex chromosomes. Females have two X chromosomes, and males have an X and a Y chromosome. These chromosomes are called X and Y because the actual chromosomes are shaped like an X or a Y (Figure 10–19). When the male sperm are developed in the testes of the male, the mature sperm will contain either an X or a Y chromosome. When the eggs are formed in the female ovaries, all will contain X chromosomes (refer to the section on meiosis in Chapter 4.) At conception, the sex chromosomes are paired together to determine the sex of the offspring. Some embryos will have an X and a Y chromosome and will be male animals. Those having two X chromosomes will be female. Obviously, the sex of the animal is determined by which sperm fertilizes the egg.

Scientists tried for many years to use the knowledge of how sex is determined to separate the X chromosome sperm from the Y chromosome sperm. Many methods were tried, and there were a lot of failures. Then, in 1989, a method for separating the two types of sperm was patented. This process is known as **sperm sexing**, or sorting. The technique makes use of a fluorescent dye that adheres to the DNA within each sperm. Since the female chromosome (the X chromosome) contains more DNA than the Y male chromosome, more dye clings to the sperm with the X chromosome. The female chromosome may contain from 2.8–7.5 percent more DNA than the male, depending on the species of animal. For example, a female cattle chromosome contains 3.8 percent more DNA than the male chromosome, and humans have about a 2.8 percent difference.

The dye-coated sperm is then passed under a laser beam that illuminates the dye. Since the female (X) chromosome gives off more light, it can be detected from the male (Y) chromosome sperm. A device called a **flow cytometer** separates the sperm into different tubes according to the amount of light given off (Figure 10–20). Through this

© ARS/USDA Photo by Scott Bauer

Figure 10–20
Through the use of a cytometer cell sorter, a laser beam illuminates dye-coated sperm and separates X and Y chromosomes.

procedure, technicians can acquire both tubes containing mostly male-producing sperm and tubes containing mostly female-producing sperm. Later developments of the method have allowed the sorting of bull semen with 85–95 percent accuracy. Other species are expected to have similar results as the procedure is perfected. Offspring from the procedure appear to be healthy and normal in all respects. Presently, all the major bull semen producers (bull studs) provide some sex-sorted semen. The process described is slow and costly, as billions sperm cells are sorted one at a time by the machine. It has been determined that semen from individual bulls may not withstand the process, while other bull semen is not so adversely affected. Because of the difficulties, the sperm cells are in a somewhat weakened state, so additional care must be taken when handling and inseminating using sex-sorted semen. Most producers recommend it be used only in virgin heifers, though as technology improves, the process will likely result in more broad application.

People have used biotechnology to breed livestock for thousands of years. The efficient reproduction of agricultural animals is essential to the survival of the industry. Over the years, such techniques as artificial insemination and embryo transfer have made the process more and more efficient with each new development and refinement. Newer technology, such as the sexing of embryos and sperm cells, promises to make the process even more efficient. Research and development will continue to be conducted in the area of livestock reproduction, and, as in the past, each new development will be a new step forward.

CHAPTER REVIEW

Student Learning Activities

1. Go to the Internet and search for companies that offer semen and embryos for sale. Choose a species and breed and research the available genetics. Make a list of all the characteristics that the breeder company lists for the particular breed. For example, for beef sires, it might be calving ease; for pigs, one characteristic may be litter size; and so on. Share your list with the class.

2. Go to the library and research the legend about the Arab raiders who began artificial insemination. Write a paper defending the legend as fact or fiction. Be sure to include the reasons why you feel it is fact or fiction. Provide evidence to support your theory.

3. Interview a producer in your area who uses embryo transfer or artificial insemination. Ask him or her why he or she uses the technology. Also ask what characteristics he or she looks for in a sire or dam.

4. What is your opinion of the claim that biotechnology in animal reproduction narrows the genetic base? Thoroughly research your ideas and give references to support your opinion.

Fill in the Blanks

1. _____ _____ is the process that involves the introduction of the male _____ into the _____ tract of a female animal by means other than the natural _____ process.

2. If animals are _____, they must go through a _____ process that requires the animals to be kept in _____ for a period of time to make sure that they do not bring _____ into the country.

3. In the _____, a process was discovered in which a protectant, such as _____, is added to the _____ before it is _____.

4. Females have to be closely _____ for signs of _____ (the period when she is ready for mating), and they also have to be _____ and _____ one at a time.

5. Processing semen involves adding _____ such as _____, egg yolk, _____, and/or _____.

6. Estrus is the time when the female allows _____, and it is controlled by the _____ and _____ of _____ at the proper time.

7. The process of _____ transfer begins with the selection of _____ _____ and _____ _____.

8. In the process of embryo transfer, the _____ animals undergo a process known as _____, causing the donor animal to release _____ eggs instead of just _____.

9. The _____ cow is made to come into estrus using a feed additive containing _____, and when the _____ _____ reaches the proper stage, the embryo is placed in the _____ of the recipient cow.

10. During sperm sorting, the _____-coated sperm is then passed under a laser beam that _____ the dye. Since the _____ (X) chromosome gives off _____ light, it can be detected from the _____ (Y) chromosome sperm.

True or False

1. Every aspect of the livestock industry is directly dependent on the reproduction of animals.

2. It is safer for the females in a herd to keep the sire on the premises than it is to use artificial insemination.

3. Artificial insemination has been used by animal producers since the Middle Ages.

4. Bull semen has been stored successfully for as long as 50 years.

5. Sperm cells of all animal species are the same shape.

6. It is beneficial for a producer to have an entire herd of females animals come into estrus at the same time.

7. Follicle-stimulating hormone causes the egg to develop within the follicle on the ovary.

8. Donor cows are usually undesirable as breeding stock.

9. Gonadotropin-releasing hormone (GnRH) and luteinizing hormone (LH) will cause a female animal to come into estrus.

10. Pregnancies resulting from embryo transfer are less likely to go full term and result in a normal calf than regular pregnancies.

Discussion

1. List the advantages of using artificial insemination.

2. Explain why the sperm preservation process developed in the 1950s was so important for the animal industry.

3. How is semen quality measured?

4. Describe the process of artificially inseminating a female animal with thawed semen.

5. Explain how a producer can synchronize the estrous cycles of a herd of females.

6. List the advantages of using embryo transfer.

7. Describe the process of embryo transfer.

8. Explain why a producer might want to have all male or all female animals.

9. Describe the process of sperm sorting.

10. What is the advantage of a dual production system?

CHAPTER 11

Biotechnology in Medicine

KEY TERMS

pathogens
pharmaceuticals
genetic diseases
biomanufacturing
pharming
vaccines
inoculation
antigens
antibodies
edible plant vaccines
gene therapy
hormone
personalized medicine
DNA fingerprints
Human Genome
    Project
markers
genomewide association
    studies (GWAS)
single nucleotide
    polymorphisms
personalized genomics
International HapMap
    Project
1000 Genomes Project
imprinting
epigenetic inheritance
organ transplant
mesenchymal stem cells
    (MSCs)
autotransplantation
xenotransplantation
xenografts

OBJECTIVES

When you have finished studying this chapter, you should be able to:

*   Explain biomanufacturing.
*   Explain the concept of pharming.
*   Discuss the role of plants and animals in human health.
*   Explain personalized medicine.
*   Define gene therapy.
*   Explain how plant pharming may reduce the cost of drug production.
*   Explain how vaccines may be administered in food crops.
*   Discuss the problems encountered in extracting hormones from animal organs.
*   List the advantages of producing drugs through milk.
*   Explain the difference between allotransplantation, autotransplantation, and xenotransplantation.

Medical Traditions

Medicines have always been a part of human life. Anthropologists and archaeologists have found ancient evidence of the uses of medicines and healing practices among early humans. Injuries, diseases, and disorders have always been a part of life, and humans have constantly searched for remedies for these problems. Traditionally, plant, animal, and mineral products have been used to help alleviate symptoms and promote healing (Figure 11–1). Although some traditional remedies were effective, most were not. In fact, many of the "cures" and remedies used by early people actually did more harm than good.

This primitive use of medicine went on for thousands of years because people simply did not know what caused the diseases or how healing from an injury took place. For example, no one knew of the existence of **pathogens**, such as viruses, bacteria, or protozoa. Notions that severe diseases such as malaria were caused by "bad night air" and the way to treat diseases was to drain blood from the patient dominated medical thinking. Often, doctors would "bleed" patients by attaching leeches or cutting into the flesh to drain away the "ill humors" that were thought to cause disease. These early approaches to medicine were founded on mistaken beliefs about the cause of disease.

It was not until the first half of the twentieth century that scientists began to truly understand the causes of diseases. Once the causes were discovered, progress was made in the way diseases and injuries were treated. Vaccines were

Figure 11–1
For many years, people have used plant and animal products to make medicines.

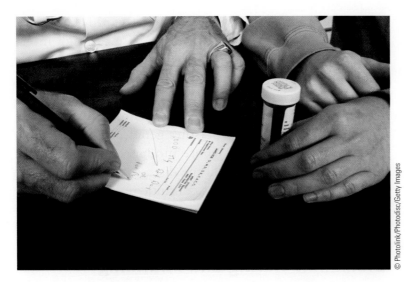

Figure 11–2
Modern medicines have been known as miracle drugs.

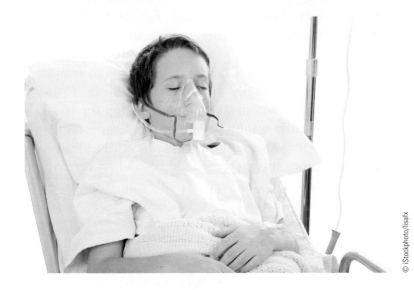

Figure 11–3
We still live with diseases for which there is no cure.

invented to prevent diseases, and appropriate treatment measures were developed to care for the sick and injured. Since the beginning of the twentieth century, many new drugs and treatments have been developed. The new **pharmaceuticals** became known as "miracle drugs" because of the miraculous effect the drugs had on the treatment of disease and infections (Figure 11–2). Antibiotics such as penicillin and its many derivatives made such diseases as tuberculosis, cholera, and plague no longer the death threat they once were.

As miraculous as modern medicines are, we still live with diseases for which there are no cures (Figure 11–3). Each year, cancer, diabetes, heart disease, and HIV/AIDS still claim millions of lives all over the world. These diseases and many chronic conditions arise through interactions between a person's genetic background and the environment. Diseases such as hemophilia, cystic fibrosis, and multiple sclerosis continue to cause suffering for lack of cures. These maladies, known as **genetic diseases**, are caused by a problem in a person's genetic makeup. In underdeveloped countries where modern medicine is not practiced on a wide scale, all the diseases that have plagued people for thousands of years continue to kill many people each year.

Biomanufacturing and Pharming

One of the greatest modern tools for treating and curing diseases is the genetic engineering of cells to produce therapeutic drugs and vaccines through **biomanufacturing**. Remember from Chapter 1 that human insulin is produced by genetically engineering bacteria to make human insulin. Through biomanufacturing, the bacteria producing human insulin are grown in large liquid cultures in a device called a bioreactor (Figure 11–4). Insulin is extracted from the bacterial cultures and processed into drugs for injection into people suffering from diabetes.

Figure 11–4

Bacteria and other cells are grown in a device called a bioreactor.

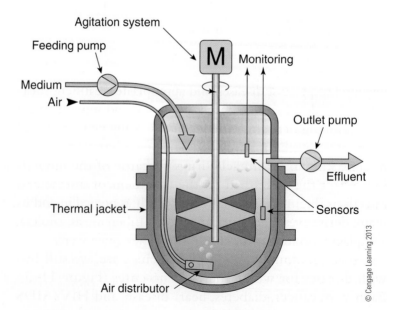

© Cengage Learning 2013

Figure11–5
The most commonly used cells to produce human medicines are cells from the ovary of the Chinese hamster.

Hormones, vaccines, therapeutic drugs, and small molecules may all be produced by genetically engineered cell cultures via biomanufacturing.

The first cells used in biomanufacturing were bacterial or fungal cells because researchers first understood how to genetically engineer and grow these cells in large cultures. Bacterial and fungal cells are still widely used in biomanufacturing to produce industrial enzymes and other useful proteins. Now, however, cell cultures used to produce human medicines are more often eukaryotic. The most commonly used mammalian cell is the Chinese hamster ovary cell because of it similarity to human cells in protein processing (Figure 11–5). Plant cells also process proteins similarly to humans.

Biomanufacturing is elegant to a relatively new yet expensive method for producing medicines. To avoid the expense of using cell cultures but keeping the benefit of using eukaryotic cells, researchers came up with the idea of **pharming**. Pharming is the production of pharmaceuticals in genetically modified whole plants or animals and gets its name from a combination of the terms "pharmaceutical" and "farming." "Pharmaceutical" refers to any type of medicinal drug or substance used to treat humans and animals. Of course, "farming" refers to the growing of agricultural plants and animals. Thus, pharming is the growing of organisms (usually genetically altered) for the purpose of producing pharmaceuticals.

Figure 11–6
Early settlers used many types of plants and herbs as medicine.

Figure 11–7
A common heart drug called digitalis comes from the foxglove plant.

Medicinal Plants and Plant Pharming

We rely on plants for our very existence. Not only do they produce oxygen, but they provide either directly or indirectly (through livestock feed) all the food we eat. The first medicines used by humans probably came from plants. We know that the roots, seeds, stems, flowers, and leaves of certain plants have medicinal value. Some medicinal plants are used to relieve pain, while others are used as a treatment or cure for diseases. Early settlers in the United States used medicinal plants, such as yellow root, ginseng, sassafras, and a wide variety of other herbs found in the wild, to help relieve their pain and treat ailments (Figure 11–6). Even today, many of our modern medicines come from plants. For example, a common heart drug called digitalis comes from the foxglove plant (Figure 11-7). Taxol, an anticancer compound, was originally derived from the bark of the Pacific yew tree.

Plant-Made Products

As mentioned previously, the use of genetically modified cells in bioreactors is widely used to produce modern medicines. Although biomanufacturing processes work quite well, input costs are high. It is expensive to build the factory, and laboratory facilities (capital costs) and bioreactor systems require large volumes of clean water, electricity, and continuous monitoring of cell culture conditions, such as pH, temperature, oxygen saturation, and nutrient levels (manufacturing costs). An economical alternative to bioreactors is the manufacture of pharmaceuticals in crop plants. With the proper genetic modification, biotech crops can make large quantities of a specific drug at a much-reduced cost. Estimates are that a 200-acre field of corn can produce the same amount of pharmaceuticals that a $400 million factory would produce in 1 year but at a fraction of the price (Figure 11–8).

Biotech crops currently in development as "pharmaceutical factories" include tobacco, corn, rice, and safflower. Some researchers prefer to use nonfood crops for pharming, such as tobacco, to avoid any chance that a crop producing a medicine could accidentally mix with a crop intended for the food supply (Figure 11–9). Another concern and

© coyote. Image from BigStockPhoto.com.

Figure 11–8
A 200-acre field of corn could produce the same amount of pharmaceuticals that a $400 million factory would produce.

© Sherjaca/www.Shutterstock.com

Figure 11–9
A concern is that pollen from a crop producing a medicine might mix with a food crop.

production challenge is that pharm crops must be prevented from pollinating nearby conventional crops of the same species or wild relatives. However, this concern about gene flow through pollination may be easily addressed. Choosing pharm crops with heavy pollen (rice or barley), planting pharm crops only in geographic areas without wild relatives, and using self-pollinating pharm crop species (soybean) or those that are unable to make pollen (male-sterile variety of potato) are all great strategies to prevent gene flow. In addition, strict U.S. Department of Agriculture field containment regulations for the growth, harvest, and storage of biotech crops make food chain contamination and "pharm gene escape" scenarios unlikely.

Land use issues have also been discussed by both supporters and critics of pharma crops. Producers, researchers, and government regulators are aware that food production must remain a priority, so developing pharma crops that will grow on marginal land is the goal. Another approach to making efficient use of arable farm land is developing crops that yield a number of usable products—perhaps a pharma-biofuel corn plant expressing a pharmaceutical specifically in the seed. In theory, once the pharmaceutical is extracted from the corn kernel, the remaining starch can be used to produce bioethanol, and the corn stover can be used to produce cellulosic ethanol. Federal regulations require dedicated harvest equipment and storage facilities for biotech crops, so a producer would need to maintain complete separation of biotech and conventional crops, if necessary.

Several pioneering companies are developing pharm crops for the production of therapeutic proteins and vaccines. Kentucky BioProcessing has a well-developed tobacco system for the production of veterinary and human medicines, including a vaccine for a disease called parvovirus in animals and a human papillomavirus vaccine (Figure 11–10.) Ventria Bioscience is using rice to produce antimicrobial proteins found in human breast milk, lysozyme, and lactoferrin. These proteins will be used in oral rehydration solutions to help

Figure 11–10
Both human and veterinary vaccines are being produced using tobacco.

prevent childhood deaths from diarrheal diseases in developing countries. SemBioSys has developed a genetically engineered safflower that produces human proinsulin in the oil seed. It is estimated that over 380 million people will have diabetes by 2025, and having an affordable supply of insulin will be critical to treat the disease. These and many more treatments and cures will be made affordable through plant pharming once strict regulatory requirements have been met.

Pharm crops lower the cost of making human medicines in genetically engineered cells. The cost savings occur as the proteins are produced. However, to extract proteins from cells, whether they are in a culture system or part of a crop plant, costs are similar. After harvest, pharm crops are taken to a Food and Drug Administration–regulated biomanufacturing facility. The crops are mechanically ground and treated with solutions to remove and purify the pharmaceutical proteins. The downstream purification, analysis, and packaging of pharmaceutical proteins is tightly regulated and must be certified as a "good manufacturing practice" process, identical to the process used by drug companies using cell culture fermentation in bioreactors.

Plant Cells for Biomanufacturing

In addition to the cost-effective, scalable whole-plant production of pharmaceuticals, plant cell cultures may also be used in closed bioreactor systems. Although similar in cost and design to mammalian, fungal, bacterial, or insect cell culture systems, plant cell cultures are advantageous in a number of ways (Figure 11–11). First, plant cell cultures are not susceptible to infection or contamination by animal, bacterial, or viral pathogens. Second, plant cells have an advantage over microbial cells because plant proteins are processed in a similar fashion to human proteins, both being higher eukaryotes. Third, plant cells seem to be good at expressing many different types of therapeutic proteins, including some that are difficult to express in mammalian cell culture. Finally, there are economic advantages to using plant cell culture versus mammalian cell culture: plant cell cultures use a simpler, cheaper nutrient media, and plant cells are easier to physically separate from the media at the end of the production process.

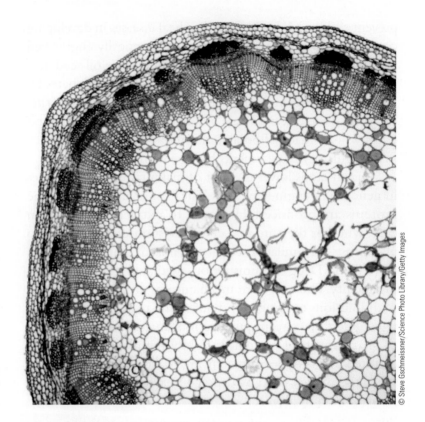

Figure11–11
Plant cells have several advantages in pharming.

© Steve Gschmeissner/Science Photo Library/Getty Images

In comparison to field-grown pharma crops, plant cell cultures have a regulatory advantage in that the process is completely contained. Like other forms of biomanufacturing, there is no risk of pharmaceutical contamination of the food supply or gene flow to related organisms. Because of the similarity to traditional biomanufacturing, regulatory approvals have been faster for the use of plant cell cultures to produce pharmaceuticals versus the use of pharm crops. Therapeutic proteins and pharmaceuticals derived from plant cell cultures are currently on the market. Protalix, an Israeli company, has developed a bioreactor system of carrot cell cultures for the cost-effective production of therapeutic proteins. The company currently produces a glucocerebrosidase enzyme used to treat Gaucher's disease.

Edible Plant Vaccines

Each year, millions of people are inoculated with **vaccines** to prevent diseases. Maladies such as measles, smallpox, and polio that once caused so much misery and death have now

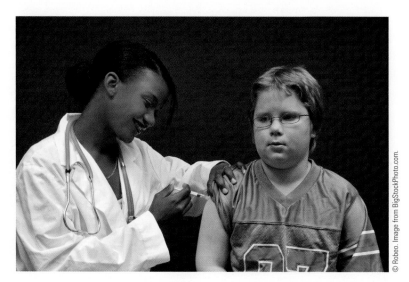

© Robeo. Image from BigStockPhoto.com.

Figure 11–12
Children are given vaccinations as
a routine part of health care. The
injection still hurts.

been practically eradicated through broad-based vaccination programs. Usually, children are given vaccination shots as a routine part of health care (Figure 11–12). In the United States, most children receive up to 24 vaccinations by the time they are 2 years old. Although the **inoculation** procedure is generally considered safe, the procedure is not without risks. Anytime a needle is inserted through the skin, there is always pain as well as a chance for infection.

In many areas of the world, children receive few if any vaccinations. Underdeveloped countries still have major disease problems that shorten the life expectancy of people. Although human relief efforts have been directed at vaccination programs, millions of people do not receive the benefit of vaccinations. Several problems exist with attempting to vaccinate populations of underdeveloped countries. First is the expense of the vaccination. Although the vaccine may be inexpensive to manufacture, the cost in transporting the vaccine and getting it to remote areas can be huge. Second is the availability of trained medical personnel to administer the shots. If the shots are not properly given, infection can occur, and/or the vaccine will not be effective. The third problem is storage of the vaccine. Most vaccines require refrigeration to prevent spoilage (Figure 11–13). Many remote areas have no electricity or other means of powering refrigeration units.

A new approach to vaccinations is genetically engineering edible plants that will provide immunity to diseases. The principle behind vaccinations is the injection of killed or

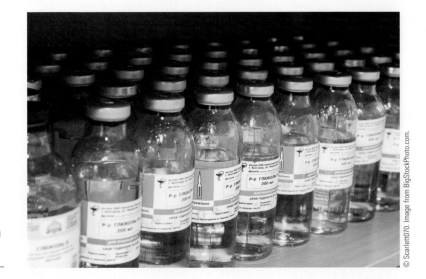

Figure 11–13
Most vaccines require refrigeration to prevent spoilage.

Figure 11–14
Scientists have successfully produced vaccines in several crops, including tomatoes.

greatly weakened disease-causing organisms, called **antigens**, into the body. The body then releases pathogen-fighting substances, called **antibodies**, which attack the disease-causing organisms. Once the antibodies have built up to a high enough level, a permanent or semipermanent resistance to the disease is created. By creating antigens within the edible part of a plant, immunity can be administered without having to inject vaccine through the skin.

Eventually, **edible plant vaccines** may play an important role in addressing vaccination challenges in developing countries. Early trials have shown that ingested antigens pass through the mucus lining of the digestive tract and impart immunity. However, the amount of fruit or vegetable eaten (dosage) must be carefully monitored to control for patient age and body size. Depending on the age of the fruit or vegetable, its degree of ripeness, and protein content, the amount of antigen in each bite will also vary. Most researchers developing edible plant vaccines plan to deal with dosage variability by processing the food into premeasured portions that will be easily stored and distributed by nonmedical personnel.

Scientists have successfully made vaccines in crops such as bananas, tomatoes, potatoes, lettuce, and rice (Figure 11–14). Focus is on developing vaccines for infectious diseases that impact large numbers of people, such as rabies, hepatitis B, measles, plague, and Norwalk virus. Researchers at the University of Tokyo are developing an oral cholera vaccine in

Figure 11–15
Bananas are considered to be the perfect food vaccines. They are a familiar food, and they are eaten raw.

genetically modified rice called MucoRice. Cholera and other diarrheal diseases kill an estimated 1.8 million people in developing countries each year, and MucoRice will provide an easily stored and distributed source of cholera vaccine. Bananas are generally considered the perfect food for vaccines (Figure 11–15). First, the fruit is widely grown in tropical areas such as West and Central Africa, Central America, and Asia, where the need for an economical and reliable means of vaccine production is great. Second, bananas are a familiar food in many developing countries, and it makes up a large percentage of the daily calorie intake. Administering vaccine-containing bananas will be easier than giving shots in many communities. Third, and perhaps most important, bananas are eaten raw and require no cooking. Heat can destroy a vaccine, so it is important that any edible plant vaccine be eaten uncooked. Since bananas are part of the everyday diet of many people in developing countries, it makes sense that they should be the food that carries the gene for vaccines. From a regulatory standpoint, commercially grown seedless bananas are vegetatively propagated and will not pose a gene flow concern. We will perhaps see this technology in widespread use in the coming years, helping to prevent many diseases and much human suffering.

Up until the past few years, genetic diseases were considered incurable, and nothing could be done in the way of prevention. Through biotechnology, scientists now have hope that these genetic disorders can be prevented or cured. It is only a matter of time before researchers are able to locate the specific genes that cause the diseases. Once the genes are located, **gene therapy** can be used as treatment. This therapy involves the insertion of genes into the human genome that corrects a problem with a person's original genes. Diseases such as Parkinson's disease, cystic fibrosis, and Down syndrome are caused by genetic problems (Figure 11–16). The idea is to use a vector, such as a virus, to insert new genetic material into the nucleus of cells involved with the genetic problem. A vector is a means to carry a disease, organism, or, in this case, genetic material. Viruses make good genetic vectors because they actually become part of an organism's DNA before the organism's immune system can overcome it. Viruses usually cause disease, but by removing the disease-causing genes from a virus, desired genetic material can be inserted. The virus then carries the new DNA into the cell nuclei of the animal that is to be genetically altered.

French medical researchers have achieved success in treating young children who because of a genetic defect were born without a functioning immune system. Such children must live inside a completely enclosed,

Figure 11–16
In the future, gene therapy may be able to prevent or cure diseases such as Down syndrome.

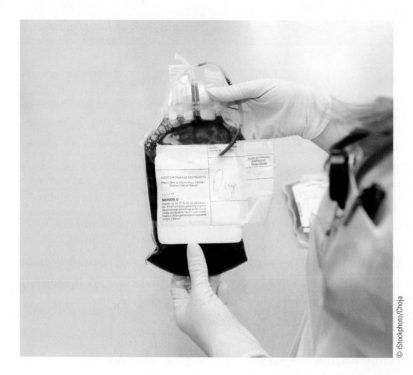

Figure 11–17
Genetically altered blood cells have been used to treat babies with a genetic defect that leaves them without an immune system.

sterile environment and seldom live very long. French scientists took a large amount of blood from a group of babies with this condition. They genetically altered the blood cells so that the cells would be able to produce blood cells that would fight infection. The genetically altered blood was then returned to the children, and the blood cells reproduced. All indications are that the treatment worked, and the children now have fully functioning immune systems (Figure 11–17). Although much of gene therapy is in the developmental stage, medical scientists hold great hope for its potential in treating genetic disorders.

Animal Pharming

The use of animal cell cultures in biomanufacturing systems is now widespread. Although the development of reproductive cloning in animals, it is now possible to genetically engineer "pharm" animals. Using animals as a source of medicine is not a new idea. In fact, animals have for many years been the source of many different types of medicines and other products of medicinal value. For example, people with disorders that require hormone replacement therapies have relied

Reading Herbal Tea Leaves: Benefits and Lore

These days, there is a lot of talk about health benefits from drinking teas.

Green, black, and oolong are considered the three major classes, and each comes from the age-old *Camellia sinensis* tea bush. But there is an even wider variety of herbal teasóinfusions derived from anything other than *C. sinensis*.

According to folklore, some herbal teas also provide benefits. But there is little clinical evidence on the effects of drinking these teas. Now, Diane McKay and Jeffrey Blumberg have looked into science-based evidence of health benefits from drinking three of the most popular herbals in America. McKay and Blumberg are with the Jean Mayer USDA Human Nutrition Research Center on Aging at Tufts University in Boston, Massachusetts. Both work at the center's Antioxidants Research Laboratory, which Blumberg directs.

One popular herbal, chamomile tea, has long been considered a soothing brew. In the early twentieth century, it was mentioned in a classic children's book about a little rabbit named Peter. At the end of a rough day, Peter's mom served him some chamomile tea. Interestingly, when Blumberg and McKay reviewed scientific literature on the bioactivity of chamomile, they found no human clinical trials that examined this calming effect.

They did, however, publish a review article on findings far beyond sedation—describing test-tube evidence that chamomile tea has moderate antioxidant and antimicrobial activities and significant antiplatelet-clumping

Antioxidants Research Laboratory scientists Diane McKay and Oliver Chen discuss the results of their hibiscus tea study, which showed the effectiveness of this beverage in reducing blood pressure. (D1814-8)

activity. Also, animal feeding studies have shown potent anti-inflammatory action and some cholesterol-lowering activity.

The researchers also published a review article describing evidence of bioactivity of peppermint tea. In test tubes, peppermint has been found to have significant antimicrobial and antiviral activities, strong antioxidant and antitumor actions, and some anti-allergenic potential. When animals were fed either moderate amounts of ground leaves or leaf extracts, researchers also noted a relaxation effect on gastrointestinal tissue and an analgesic and anesthetic effect in the nervous system.

The researchers found several human studies involving peppermint oil, but they found no data from human clinical trials involving peppermint tea. McKay and Blumberg have concluded that the available research on herbal teas is compelling enough to suggest clinical studies.

McKay has led a human clinical trial to test whether drinking hibiscus tea affects blood pressure. She tested 65 volunteers, aged 30–70 years, who were pre- or mildly hypertensive. Blood pressure readings of 120/80 or greater are considered a risk factor for heart disease, stroke, and kidney disease.

For 6 weeks, about half the group was randomly selected to drink three cups of hibiscus tea daily. The others drank a placebo beverage containing artificial hibiscus flavoring and color. All participants were advised to follow their usual diet and maintain their normal level of activity. Before the start of the study, blood pressure was measured twice—1 week apart—and at weekly intervals thereafter.

The findings show that the volunteers who drank hibiscus tea had a 7.2-point drop in their systolic blood pressure (the top number), and those who drank the placebo beverage had a 1.3-point drop.

In a subgroup analysis of the 30 volunteers who had the highest systolic blood pressure readings (129 or above) overall at the start of the study, those assigned to drink hibiscus tea showed the greatest response to hibiscus tea drinking. Their systolic blood pressure went down by 13.2 points, diastolic blood pressure went down by 6.4 points, and mean arterial pressure went down by 8.7 points.

on animal sources of hormones for treatment in the recent past. A **hormone** is a chemical substance secreted by an organ that has important regulatory effects on other behavior of cells elsewhere in the organism. Hormones, like insulin used to treat diabetes in humans, used to come from organs of animals slaughtered for meat However, traditional processes of extracting hormones from animal organs are slow and expensive. These methods have been widely replaced by biomanufacturing.

One of the greatest promises for efficiency producing drugs is to create them in milk. Nature already uses milk as a means of conveying immunity to young mammals through the colostrum that the mother gives for the first few days after young are born (Figure 11–18). Disease immunity, nutrients, and digestive clearing agents are all present in the cow's milk immediately following birth. This principle can be carried further through the use of genetic engineering. Cows can be implanted with embryos that have been genetically engineered to possess a gene that produces a certain drug through the mammary gland. Once the new animal has been born and is raised to maturity, the milk from the animal will contain the desired drug. Since the transferred gene is activated only in the mammary gland, the rest of the animal's body is perfectly normal, and the animal continues to produce a constant supply of the drug-containing milk. Drugs, such as human serum albumin, that are given to

Figure11–18
Milk holds great promise to produce drugs. Nature uses milk to impart immunities to newborn animals.

© Mark Harrison/www.Shutterstock.com

patients with blood loss; lactoferrin that is used to combat digestive disorders in human infants; and human insulin are all examples of drugs that can be successfully produced in milk from genetically modified cows.

Scientists have genetically engineered rabbits, goats, sheep, pigs, and cows to produce transgenic proteins in their milk. Table 11–1 includes many of the milk-derived pharmaceuticals currently in development or on the market. Many of the milk-derived complex pharmaceuticals are blood proteins or monoclonal antibodies. These proteins are used to treat blood coagulation diseases, interact with the immune system proteins to help treat autoimmune diseases like rheumatoid arthritis and lupus, or target cancer cells in diseases like B-cell leukemia and non-Hodgkin's lymphoma.

TABLE 11–1 Pharming Products Currently in Development

Animal	Drug/Protein	Use
Sheep	Alpha 1 antitrypsin	Deficiency leads to emphysema
Sheep	CFTR	Treatment of cystic fibrosis
Sheep	Tissue plasminogen activator	Treatment of thrombosis
Sheep	Factor VIII, IX	Treatment of hemophilia
Sheep	Fibrinogen	Treatment of wound healing
Pig	Tissue plasminogen activator	Treatment of thrombosis
Pig	Factor VIII, IX	Treatment of hemophilia
Goat	Human protein C	Treatment of thrombosis
Goat	Antithrombin 3	Treatment of thrombosis
Goat	Glutamic acid decarboxylase	Treatment of type 1 diabetes
Goat	Pro542	Treatment of HIV
Cow	Alpha-lactalbumin	Anti-infection
Cow	Factor VIII	Treatment of hemophilia
Cow	Fibrinogen	Wound healing
Cow	Collagen I, collagen II	Tissue repair, treatment of rheumatoid arthritis
Cow	Lactoferrin	Treatment of gastrointestinal tract infection and infectious arthritis
Cow	Human serum albumin	Maintenance of blood volume
Chicken, cow, goat	Monoclonal antibodies	Other vaccine production

Source: Courtesy of Animal and Plant Health Inspection Service, (APHIS) U.S. Department of Agriculture.

Beside milk, other animal sources of pharmaceutical proteins and vaccines are blood and urine. These fluids have the advantage of being usable as soon as the genetically modified animals are born, whereas deriving drugs from animal milk is delayed until the animal matures, reproduces, and begins to give milk. Researchers have also had recent successes with the development of genetically modified chickens that produce active pharmaceutical proteins in their eggs.

Personalized Medicine

Whether we choose to use biomanufacturing or pharming to produce therapeutic drugs, the first step in designing pharmaceuticals is understanding the cause of disease. In figuring out the biological causes of disease, doctors and research scientists have a new weapon in their arsenal: genome analysis. Correlating human genome variations with disease may allow for the development of diagnostic tests and personalized drug treatments. **Personalized medicine** is based on our understanding of the individuality of the genetic code. Since the 1980s, we have had the technology to generate **DNA fingerprints** for individuals. These fingerprints represent regions of the human genome that vary in length between different people and are used in criminal forensics, paternity testing, and identification of human remains. However, the genomic regions used in DNA fingerprinting do not include protein-coding genes and are not useful for identifying the causes of genetic diseases. Because of the **Human Genome Project**, wealth of information on the size, organization, and content of the human genome has been made available to the medical community (Figure 11–19). Researchers have been working to identify specific genes and nearby genome variations, or **markers**, that correlate with medically important traits, such as susceptibility to a genetic disease or the ability to metabolize a certain type of drug. These projects are called **genome-wide association studies (GWAS)**. Typically, the genome variations examined in a GWAS are **single nucleotide polymorphisms (SNPs)**, which are single base changes in the DNA that may vary between people. To date, nearly 600 GWAS studies have found approximately 800 SNPs correlating to medical traits, such as age-related macular degeneration, Crohn's disease,

Figure11–19
The Human Genome Project has given the medical community a wealth of information.

and response to hepatitis C drug therapy. Correlating variations in human DNA sequences with medical traits gives researchers a starting point for uncovering the biological causes of disease, which will hopefully lead to more effective drug treatments.

Some private companies are now offering SNP testing directly to consumers. For example, 23andMe is a company that offers to genotype 600,000 SNPs for paying customers and to make statistical predictions for relative disease risks. Disease risk predictions are somewhat controversial, as they rely on the statistical power of currently published GWAS. As more genomes are sequenced, the ability to use a particular SNP or set of SNPs to predict risk of disease will increase. Companies like 23andMe also offer ethnic ancestry predictions based on SNP analysis, which is appealing to customers interested in exploring family origins.

In addition to GWAS, which look at a fraction of the human genome, improvements in genome sequencing technology has made whole-genome sequencing a possibility for medical clinics. With the cost of sequencing an entire 3-billion-base-pair human genome now approaching $1000, we are entering the age of **personalized genomics**. Medical researchers are beginning to choose whole-genome sequencing for patients with unusual genetic diseases in order to get the maximum amount of biological information to understand and treat disease causes (Figure 11–20).

International cooperation between research groups has led to the **International HapMap Project**, an effort to characterize genomic similarities and differences between various groups of people. The project generated a publicly available database of SNPs compiled from four populations, including African, Asian, and European peoples. Currently, the governments of the United States, the United Kingdom, China, and Germany are working on the **1000 Genomes Project** to expand our knowledge of human genomic variation beyond SNPs into protein-coding regions and other parts of the human genome. By cataloging human genomic variation, these research groups are providing data that may be used to support medical research into the underlying causes of disease and to identify the best treatments for individual patients based on their genetic background.

As we move closer to personal knowledge of our own genes and genomes, we aim to develop personalized medical interventions and treatments that are based on a biological understanding of disease. However, knowledge of our genome sequence is only one part of the solution. We must also understand how our genome interacts with the environment, and it turns out that the environment can play an important role in determining which parts of the genome are "turned on" in cells. The phenomenon of mammalian **imprinting** has been observed for some time; certain traits will be expressed in offspring only if they are inherited from either the maternal or the paternal parent. We now understand that imprinting is an example of **epigenetic inheritance**, which may be defined as an observed change in phenotype or underlying gene expression without a corresponding change in the DNA sequence. Epigenetics has been best studied during cell differentiation, when genes not needed for a specific cell's function are silenced by methylation of the DNA or by other chemical modification of nearby chromosomal proteins.

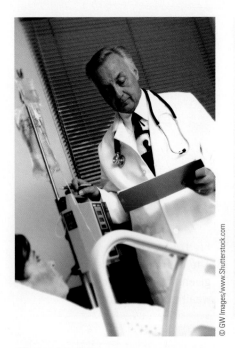

Figure 11–20
Medical researchers are beginning to choose whole-genome sequencing for patients with unusual genetic diseases.

© GW Images/www.Shutterstock.com

Stem Cells and Regenerative Medicine

Human disease and injury sometimes leads to the breakdown of vital tissues and organs. The concept of using "spare parts" to replace worn items in the body has been around for a long time.

The Production of Transplant Organs

For many years, stories of science fiction have been written about replacing damaged or diseased human organs with healthy ones. In 1818, a novel by a British author named Mary Shelly was published. The plot of the story centered around a young medical student named Frankenstein, who constructed an entire human body by using organs and other body parts from the bodies of dead people. As the story goes, the creation was given life and became known as Frankenstein's monster. As most of us know from reading the novel and watching one of the many movies developed around the plot, a monster was created that ultimately killed his creator and had to be destroyed. Since that time, many scientists have dreamed of using healthy organs from fatal accident victims to replace nonfunctioning organs in living humans.

In the 1960s, this vision came true when a heart from one person was successfully transplanted into the chest of another person. Although the patient lived for only a few days, it proved that **organ transplant** could be done. Today, the transplantation of hearts, lungs, kidneys, and other organs has become routine. Each year, there are over 2,000 heart transplant operations performed in the United States alone (Figure 11–21). In addition, there are tens of thousands of people awaiting transplant organs at any given time. Although many people indicate their wish to be an organ donor if they are killed in an accident, there are many more potential recipients than donors.

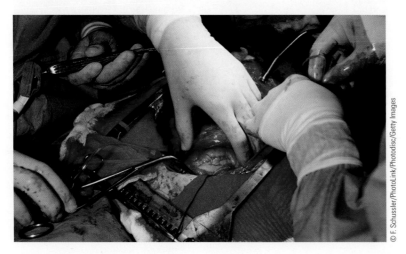

© F. Schussler/PhotoLink/Photodisc/Getty Images

Figure11–21
Each year, there are over 2,000 heart transplant operations in the United States.

Traditional organ transplantation poses two major challenges. The first challenge is the shortage of organs to transplant. Surgeons have only a very short time to remove organs from accident victims before the organs are no longer usable. Since there are so many people in need of an organ transplant and there are so few organs donated, many people die each year waiting for a usable organ. The second major challenge is that the body naturally rejects "non-self" tissue. The immune system is effective because it has a means of recognizing any type of foreign tissue or organism that invades the body (Figure 11–22). Immune system recognition triggers an immune response that immediately attacks the invading tissue or organism as a means of preventing disease. Almost all the organ transplant failures have been attributed to tissue rejection, and transplant recipients must take medicine to suppress their natural immune responses, and this may pose other health problems.

Research scientists have made a lot of progress in preventing organ rejection, primarily by matching genetically similar organ donors and recipients. Ideally, a patient would do best with a genetically identical organ donor. And barring the presence of a selfless identical twin, what could be more genetically similar to a patient in need than his or her very own cells? Recent advances in regenerative medicine allow for the use of adult **mesenchymal stem cells (MSCs)** taken from healthy areas of the patient's body to be used in repairing diseased and damaged tissues. This is known as **autotransplantation**.

Figure 11–22

The immune system recognizes and attacks any type of foreign tissue or organism that invades the body.

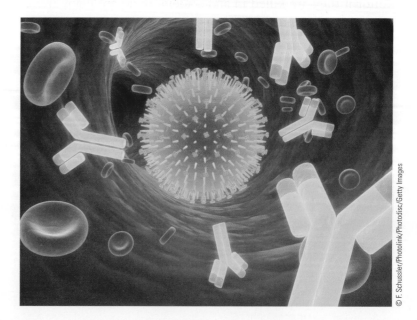

MSCs are found in the bone marrow, fat, umbilical cord blood, amniotic fluid, placental tissue, amniotic fluid, dental pulp, and a number of other human tissues. MSCs are pluripotent, meaning that they have the potential to repair and develop into many different types of tissue, including bone, muscle, and neurons. These cells are truly amazing, naturally acting as the "paramedics of the body" and mobilizing from the bone marrow and traveling to sites of tissue injury to aid in repair. Scientists are actively working to understand the natural healing activity of MSCs and to use this ability in regenerative medical therapies.

In addition to MSCs, there are other types of adult stem cells that may be used to regenerate tissues, such as hematopoietic stem cells (HSCs), which develop into blood cells. Bone marrow allotransplants to treat leukemias, lymphomas, and other blood diseases have been common for years. HSCs may be obtained from umbilical cord blood, and some parents are now choosing to bank their baby's cord blood at birth with the reasoning that cord blood may be needed for stem cell autotransplantation later in the child's life (Figure 11–23). In another form of autotransplantation, a portion of an adult patient's own bone marrow may be removed and subjected to optical cell sorting. Using a method called flow cytometry, lasers can "read" the surface proteins of cells, sorting healthy cells from diseased cells. Healthy cells, including MSCs and HSCs, may then be returned to the patient's body after the patient undergoes treatment, like chemotherapy or radiation, to kill cancerous or diseased cells.

The use of stem cells in regenerative medicine will have a tremendous impact on the ability of doctors to repair tissues and treat disease. There is controversy around stem cell technology because of concerns regarding the use of human embryos to derive embryonic stem cells for research and medical applications. Since those times, research focus has shifted to the use of adult stem cells in large part because patients will be guaranteed a genetic tissue match if their own cells are used in a medical treatment (Figure 11–24). Embryonic stem cells are arguably more useful than adult stem cells in the sense that they may develop into many types of tissue, but they have the drawback of being a form of "autotransplant" and may be recognized by a patient's immune system in a manner similar to donor-derived organs. In addition, the controversy over the use of human embryos will remain an issue.

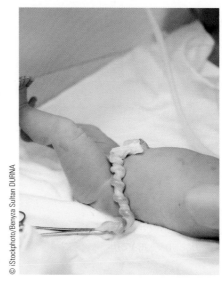

© iStockphoto/Benyza Sultan DURNA

Figure 11–23
Parents sometimes preserve a newborn's umbilical cord because it contains stem cells that could be used later in the child's life.

Figure 11–24
Adult stem cells can be cultured for use in research.

Xenotransplantation

In the absence of genetically similar human organ donors (allotransplant) or an appropriate adult stem cell treatment (autotransplant), researchers have also been working on the possibility of organ transplants from other species, known as **xenotransplantation**. Since human and animal DNA differ, conventional transplants would be difficult using animal organs. However, through the use of gene transfer, using animals as organ donors may someday be possible. If this process is developed, it could solve the problems of organ unavailability and rejection. Pigs are considered to be the most likely choice for growing organs, as the vital organs of pigs are about the same size as human organs (Figure 11–25). To make xenotransplantation a successful alternative, however, pigs will likely be genetically modified to display human proteins on their cell surfaces rather than pig proteins. The idea behind the concept is to transfer enough human DNA into pig embryos to cause the recipient to accept the transplanted organ without rejection. In fact, DNA from the recipient patient could be used for the genetic modification to ensure a genetic match. Also, there is some concern that naturally occurring porcine viruses and other pathogens may affect human xenotransplantation recipients. Once a doctor identified a person who may eventually need a transplant, the genetic transfer could take place, and an animal with the

© Life on White. Image from BigStockPhoto.com.

Figure 11–25
Pigs are considered to be the most likely animals for growing organs for humans. Their organs are about the same size as human organs and their systems are similar.

recipient's DNA could be grown and ready within 6 months. Another possibility is that the gene that controls DNA rejection could be identified and "turned off."

The use of animal replacement parts or **xenografts** is already a reality. For many years, valves from pig's hearts have been used to replace heart valves in humans. Faulty heart valves often occur in people, especially as they begin to age, and replacement may be the only medical alternative. Researchers have tried to use replacement valves, but they have encountered two problems. One is that mechanical valves tend to damage the blood cells as they travel in the heart and through the valve. The other problem is that the mechanical valve tends to accumulate blood residue, and this problem leads to the valve sticking, which in turn can cause death. Pig valves are treated in a special process that essentially renders them inanimate yet allows them to retain the properties that make them functional. This process prevents the rejection of the valve by the body (Figure 11–26). Very thin layers of skin from pigs have also been used for years to treat severe burns in humans. Researchers in the field of xenotransplantation are working to address tissue compatibility issues and predict that we will eventually be able to grow human replacement organs in genetically modified pigs.

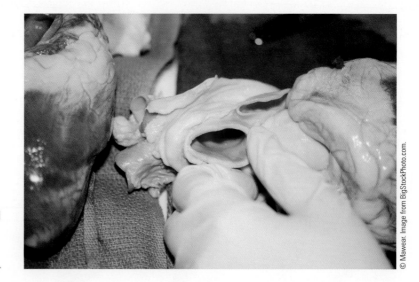

Figure 11–26
Pig valves are specially treated and used as replacements for human heart valves.

Summary

Biotechnology and modern medicine are tightly linked. From the traditional use of plants and animals as a source of medicines, we have evolved pharming technologies that allow the genetic modification of plants and animals to produce cost-effective medicines and vaccines. Thanks to the genomics revolution, we have entered the age of personalized medicine and gene therapy. We know more than ever about the underlying causes of human disease and can use this knowledge to custom-design effective new drugs and gene therapies. In the field of organ transplants and regenerative medicine, use of adult stem cells has great potential to revolutionize our ability to repair the human body. To decrease the likelihood of a negative immune response to repair tissues, patients may donate adult stem cells to themselves (cell sorting or cord blood banking). Someday, entire organs may be grown in genetically modified pigs and xenotransplanted to patients in need of a donation. The use of biotechnology promises to revolutionize the way we prevent and treat diseases, but much work remains to be done. The next few years should see the movement of many targeted medical therapies, from the research laboratory into the clinic, with the potential to alleviate much human suffering.

CHAPTER REVIEW

Student Learning Activities

1. Choose an underdeveloped country and research the problems in the country regarding health issues. What diseases cause the most problems? Is there a vaccine for the particular diseases? Write a paragraph on your perception of how edible plant vaccines might work in the country. What foods might be used?

2. Locate information on one of the genetic diseases that affects humans. Write a report about the latest research in the development of gene therapy treatment for the disease. What is the outlook for the future regarding treatment? How is the treatment expected to be administered? Share your report with the class.

3. Conduct an Internet search using the word "pharming." Choose a plant currently under development as a means of producing a pharmaceutical. Report to the class about the plant or crop. Include in your report such information as how the plant is genetically altered, what genes are inserted, what pharmaceutical is produced, and the use for the drug. Also include information on the status of the plant. Has it been approved for production? If not, when will it be approved?

4. Interview a medical doctor or a pharmacist to determine his or her ideas about pharming. What does he or she consider to be the greatest benefit and/or potential breakthrough in this area?

Fill in the Blanks

1. Traditionally, _____, _____, and _____ products have been used to help alleviate _____ and promote healing.

2. One of the greatest modern tools for treating and curing diseases is _____ _____ of cells to produce therapeutic _____ and _____ through _____.

3. Pharming is the production of pharmaceuticals in genetically _____ whole _____ or _____ and gets its name from a combination of the terms "_____" and "_____."

4. The principle behind _____ is the injection of _____ or greatly _____ disease-causing organisms, called _____, into the body.

5. _____ therapy involves the _____ of genes into the _____ genome that _____ a problem with a person's original genes.

6. A _____ is a means to carry a _____, organism, or _____ material.

7. A _____ is a chemical substance _____ by an organ that has important _____ effects on other _____ of cells elsewhere in the organism.

8. Correlating human _____ variations with _____ may allow for the development of _____ tests and _____ drug treatments.

9. DNA _____ represent regions of the human genome that vary in _____ between different people and are used in _____ _____, _____, and identification of human remains.

10. International cooperation between research groups has lead to the _____ _____ _____, an effort to characterize genomic _____ and _____ between various groups of people.

True or False . . . .

1. Diseases that are easily treatable with antibiotics, such as cholera, no longer kill humans.

2. Pharming is used because it is cheaper than biomanufacturing.

3. All researchers prefer to use food crops for plant pharming.

4. Regulatory approvals have been faster for the use of pharmaceutical crops to produce pharmaceuticals versus the use of plant cell cultures.

5. Edible plant vaccines work because ingested antigens pass through the mucous lining of the digestive tract and impart immunity.

6. Rice is considered the perfect food for vaccines.

7. The advantage of using animal blood or urine for the production of drugs and vaccines is that these fluids can be used immediately on the birth of the animal.

8. Single nucleotide polymorphisms are single base changes in the DNA that may vary between people.

9. Almost all organ transplant failures are due to tissue rejection by the recipient's immune system.

10. The best replacement for damaged human heart valves are mechanical valves.

Discussion . . . .

1. What is biomanufacturing?

2. Why are researchers concerned with gene flow in plants used for pharming? How can they prevent gene flow?

3. What are the advantages of using plant cell cultures rather than animal or bacterial cell cultures for biomanufacturing?

4. What are the barriers to providing vaccinations to the populations of underdeveloped countries?

5. Explain how vaccinations work.

6. Describe the main problem with using edible plant vaccines.

7. Explain how milk can be used as a drug manufacturing and delivery system.

8. Explain the importance of genome-wide association studies (GWAS).

9. What is epigenetic inheritance?

10. Identify and describe the three main types of organ transplant.

CHAPTER 12

Biotechnology in the Food Industry

KEY TERMS
dietary deficiencies
proteins
carbohydrates
vitamins
minerals
phytochemicals
vitamin A
anemia
beta carotene
biofortified
"Golden Rice"
legumes
enzymes
drying
canning
mechanical refrigeration
freeze-drying
rennin
curds
whey
chymosin
biomanufacturing

OBJECTIVES

When you have finished studying this chapter, you should be able to:

- Explain the importance of proper nutrition.
- Define macronutrients and micronutrients.
- Discuss the importance of vitamin A in the human diet.
- Explain why so many people in developing countries suffer from vitamin A deficiency.
- List common sources of vitamin A.
- Define "Golden Rice."
- Discuss how genetic engineering is used to produce biofortified crops.
- Describe the problems people have had in preserving foods.
- Trace some of the important developments in preserving foods.
- Explain why various methods of food preserving work.
- Define enzymes.
- List some of the functions of enzymes in the human body.
- Explain how enzymes are used in the food processing industry.
- Describe how genetic engineering is used to produce enzymes through biomanufacturing.

Figure 12-1
It is estimated that over 800 million people are not receiving adequate nutrition.

Meeting Dietary Needs

Just like the water we drink and the air we breathe, food is the very basis of life. Without the proper daily intake of the right types of food, our bodies cannot function as they should. For many years, people living in the United States have had access to an abundance of highly nutritional, safe-to-eat foods. This is not true in much of the rest of the world. In many countries, people struggle to find enough food to eat, and millions of them go hungry or do not have the proper intake of dietary macronutrients (proteins, carbohydrates, and lipids) or micronutrients (vitamins, minerals, and phytochemicals). Current global food production meets the needs of almost 7 billion people. Estimates are that over 800 million people are not receiving adequate nutrition in their diets (Figure 12–1). By 2050, we will have needed to increase current food production by about 50 percent to meet the nutritional needs of a projected global population of 10 billion.

Dietary deficiencies cause more problems than just that of hunger. Our bodies need a wide variety of different nutrients to perform such functions as cell replacement, the development and maintenance of our immune systems, and the generation of the energy we need in our daily lives. Macronutrients are consumed by people in large quantities and provide most of our energy and building blocks for growth. **Proteins** provide the amino acid building blocks needed for creating new cells and tissues. **Carbohydrates** are used by cells to create energy. Lipids (fats and oils) are needed for a number of important roles, including absorption of vitamins A, D, E, and K; the maintenance of a healthy immune system; storage of energy; insulation of organs; regulation of body temperature; and brain development. Micronutrients are needed by the body in very small quantities, usually less than 100 mg per day. **Vitamins** aid the immune system and assist in the function of enzymes. **Minerals**, such as iron, zinc, and iodine, are used to build bones, blood, and other tissues as well as to help the immune system to function properly. **Phytochemicals** are plant-derived molecules that may have anti-inflammatory, antioxidant, or immune-enhancing properties. It is known that a diet rich in fruits and vegetables promotes health, and scientists are currently investigating how plant-based foods might be responsible for positive health effects.

A shortage of either macro- or micronutrients can cause our bodies to malfunction, which in turn leads to disease or even death. For example, the World Health Organization estimates that over 100 million children in the world suffer from a deficiency of **vitamin A**. This deficiency is so severe that it causes blindness and eventual death. The lack of iron in the diet causes **anemia**, a condition suffered by over 2 billion people in the world.

Results of vitamin A and iron deficiencies are most notable in such countries as India, Bangladesh, the Philippines, Vietnam, Indonesia, and Thailand, all of which use rice as a main part of their diet. In fact, rice makes up over 80 percent of the world's diet (Figure 12–2). Rice is so popular as a cereal crop because it is easily grown in countries where there are large populations and also because the grain is stable when stored and does not spoil very quickly.

While it is a good source of energy, rice is devoid of vitamin A and iron. When almost an entire diet is made up of rice, a severe vitamin A deficiency, along with other nutritional deficiencies, results. Vitamin A may be made in the human body by metabolizing the phytochemical **beta-carotene**, which provides the yellow or orange color to fruits and vegetables (Figure 12–3). Green leafy vegetables, carrots, and some types of fruits are good sources of vitamin A. However, most of the poor people in underdeveloped countries do not consume enough of these plant products

Figure 12–2
Rice is the main part of the diet for much of the world's population.

Figure 12–3
Beta carotene is found in orange vegetables, such as carrots.

Figure 12–4
Scientists have successfully spliced a beta-carotene–producing gene from daffodils into the genome of rice.

to receive an adequate amount of beta-carotene. Fruits and vegetables are much more expensive than rice, so most of the poorer people rely on rice to provide their daily diets.

To address vitamin A deficiency in developing countries, scientists Ingo Potrykus and Peter Beyer came up with a plan to design rice with nutritional levels of beta-carotene. This idea became a reality in the late 1990s when scientists were successful in placing a beta-carotene–generating gene from a daffodil plant into rice. Daffodils are ornamental plants that produce large yellow flowers (Figure 12–4). The nutritionally improved or **biofortified** crop is known as Golden Rice because of its yellow-orange color. Traditionally grown rice varieties do not have the complete set of genes needed to produce beta-carotene, so Golden Rice was genetically engineered to have the needed genes.

Golden Rice has been in development since the 1990s, has been in field trials since 2004, and is forecasted to reach farmers' fields by 2012. Along the way, early Golden Rice strains were criticized for not containing enough beta-carotene to make a nutritional impact. Current versions of Golden Rice contain about 37 times more beta-carotene than early versions and will have the desired positive impact on vitamin A deficiency in regions where rice is a dietary staple (Figure 12–5). In addition to beta-carotene, Golden Rice researchers are now working on increasing levels of vitamin E, iron, zinc, and protein in this biotech crop.

Unlike the majority of biotech crops, Golden Rice was developed for purely humanitarian purposes. The inventors' goal was to alleviate suffering caused by nutritional deficiencies among the poor people of the world. To achieve this goal, Golden Rice seed will be given to producers in developing countries (earning less than $10,000 per year). Golden Rice strains derived from local cultivars have been developed, allowing farmers in developing countries to grow rice suited to their field conditions. Farmers will be able to locally grow, harvest, sell, and save Golden Rice as they would a traditional, nonbiotech variety. By providing the seed and technology without cost to the farmers, the goal of alleviating vitamin A deficiency in rice-dependent communities will be achieved.

In sub-Saharan Africa and other developing regions, cassava is an important staple crop and, like rice, may make up the bulk of local diets. Scientists at the Donald Danforth Plant Science Center in St. Louis, Missouri, have been working on the BioCassava Plus program since 2005. They have genetically engineered a cassava with increased levels of beta-carotene, protein, iron, zinc, and vitamin E (Figure 12–6). Field trials in Nigeria began in early 2009, and this crop promises to improve the nutrition of approximately 800 million people who rely on cassava as a staple. Other genetically engineered biofortified crops in the pipeline

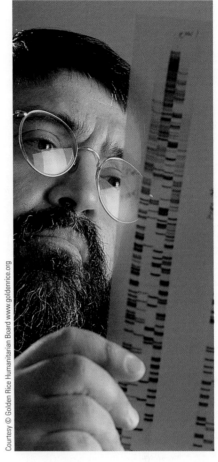

Courtesy © Golden Rice Humanitarian Board www.goldenrice.org

Figure 12–5
Current versions of Golden Rice contain about 37 times more beta-carotene than earlier versions. Golden Rice (right) has a golden color. Conventional rice is white (left).

© vtupinamba. Image from BigStockPhoto.com.

Figure 12–6
Cassava has been genetically engineered to have higher levels of beta-carotene, protein, zinc, iron, and vitamin E.

Figure 12–7

Legumes, such as beans, are used as a source of protein. Plant protein is generally lower quality than protein from animal sources.

include maize with increased beta-carotene, folate, and ascorbate (vitamin C); lettuce and carrots with increased calcium; soybeans with high levels of omega-3 fatty acids; rice with increased bioavailable iron; and potatoes with higher levels of protein.

Biofortification of crops often focuses on increasing levels of micronutrients, such as vitamins or minerals. However, people in developing countries that rely heavily on starchy staple crops are often lacking in the macronutrient protein. Traditional sources of protein include animal products, which are often unaffordable, and legumes. **Legumes** are plants that have a symbiotic relationship with soil bacteria that live in nodules along the plant root. The legume provides a home for the bacteria and some nutrition, while the bacteria take up nitrogen from the atmosphere and convert it into a form usable by the plant. Nitrogen is needed to make amino acids, which are the building blocks of proteins. As a result of this symbiosis, legumes, such as soybeans, peanuts, peas, and beans, have seeds with relatively high protein content (Figure 12–7). In some developing countries, legumes are also too expensive for the poorest residents to include in their daily diet. For these areas, genetically engineering high-protein staple crops will be an important goal.

Biotech scientists have the ability to custom design food crops with specific levels of macro- and micronutrients. The process is highly regulated and takes years of careful work in the laboratory and in the field. The resulting crops, however, have the potential to sustain millions of hungry and malnourished people.

Preserving Foods through Biotechnology

Perhaps the most important responsibility of the agriculture industry is to provide a safe, abundant food supply, yet there are many steps on the way from farm to fork. With the improved crops discussed in Chapter 9 and biofortified crops described in the previous section, recent biotechnology advances in increasing food production and nutritional value have been discussed. Over the centuries, early forms of biotechnology played a crucial role in the preservation of food crops and animals by providing the technology to properly preserve food until it can reach our tables for consumption.

Figure 12–8
Plant seed, such as grain, does not spoil very easily because the seed dries as it matures.

In the not-so-distant past, people were unable to store much food. As soon as animals were slaughtered, the meat and other products harvested from the animal began to deteriorate. Plant products were also difficult to store, especially fresh fruits. Plant seed such as grain usually did not spoil very easily because the seed dried in the sun as it matured (Figure 12–8). It did little good to have a bountiful harvest or be able to slaughter a large animal if the food could not be preserved very long. Large feasts were followed by times when food was in short supply and people went hungry. In the warm months, people could eat fresh fruits and vegetables, but in the colder months, they relied on stored grain to supplement the meat in their diet.

Of course, people had no idea what caused food to go bad. Only within the past 150 years have we discovered that microbes such as bacteria and fungus spores were responsible for breaking down the tissues in the food material (Figure 12–9).

Also, we have found out that as soon as a fruit or vegetable is picked, **enzymes** go to work to change the composition of the food. Enzyme action can render many types of food unpalatable or maybe even inedible. The whole key to preserving food was to discover ways of stopping microbial action and at the same time preserving the flavor and nutrition of the food. This usually involves killing the microbes and/or preventing them from multiplying.

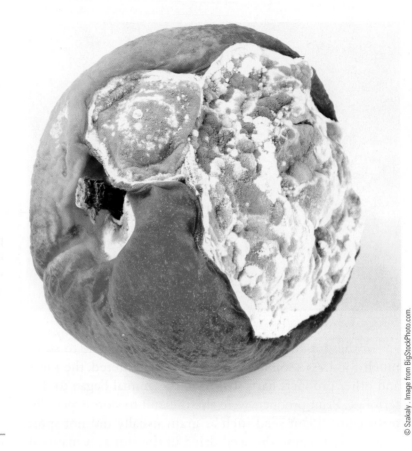

Figure 12–9
Microbes such as bacteria and
fungus spores break down the
tissue in food material.

© Szakaly . Image from BigStockPhoto.com.

Drying and Salting

Although at first they did not understand the causes of food
spoilage, people began to look for ways in which food could
be stored for future use. It is highly likely that they observed
that dried seeds did not spoil and reasoned that if they dried
foods, they might also be stored. Through experimentation,
they learned to preserve meats, fruits, and fish by **drying** them
in the sun or over a fire. This method worked because the
drying process removed the moisture in the food necessary
for bacterial growth.

People learned that salt also provided a way of preventing
food spoilage. By packing it in salt or by storing it in a brine
solution, food could be kept edible for extended times. The
action of the salt pulls moisture from the bacteria through
osmosis, and this effectively kills food-spoiling microorganisms. Over the ages, salt became a precious commodity that
was a stable part of many commercial activities. This use of

Figure 12–10
People still like the taste of dried foods such as sausage.

drying, salting, and pickling as a means of preventing spoilage is still used because people like the taste of the foods preserved in this manner (Figure 12–10).

Canning

One of the greatest of all advances in food preservation is **canning**. This process was developed around the turn of the nineteenth century when French Emperor Napoleon offered a large financial reward to anyone who could develop a way to preserve food so that troops in the field could receive enough food delivered over long distances (Figure 12–11). A Frenchman by the name of Nicolas Appert won the award by preserving peas in a way that would keep them indefinitely. He first cooked peas and then placed them in wine bottles. The bottles were then heated to a high temperature and sealed while they were hot. The results were phenomenal in that large quantities of peas and other foods could be transported over long distances over periods of many weeks and arrive in edible condition. This began the industry of canning.

Although the process was effective and revolutionized food preservation, neither Appert nor anyone else understood why it worked. The method worked because the heat destroyed the microorganisms in the food. When the water started to turn to steam, air was driven out of the

Figure 12–11
Preserving food by canning was a result of the emperor Napoleon's need to feed his army in the field.

Figure 12–12
Canning has evolved over time with such advancements as the use of safe, nontoxic cans.

bottles. When a seal was applied, no other microorganisms could get into the bottle to cause the food to spoil. Canning evolved over time with advancements such as the use of cans (Figure 12–12).

Cans are superior to bottles or glass jars because they do not break and can withstand transportation better. One major problem with cans was unknown at the time. The first cans were sealed with lead because the soft metal was readily available and could be melted at a relatively low temperature. These characteristics seemed to make lead the perfect substance with which to seal food cans. However, as was discovered many years later, lead is a poison that slowly builds up in the body and causes severe health problems. Modern cans are now sealed with nontoxic materials.

Another problem that was overcome was that even though the food was processed at a high temperature, sometimes microorganisms survived and caused the food to spoil. The cans could be heated to only a limited degree because the water in the sealed cans would turn to steam, and the rapidly expanding steam would explode the cans. This problem was solved by sealing the cans (or glass jars) in the conventional way by creating steam and sealing the cans before air could get in. Once the cans of food were sealed, they were put under pressure and heated further. This method makes use of the laws of physics involving the temperature at which water boils. Water converts from a liquid to steam at 212°F (100°C) at atmospheric pressure. If water is put under pressure, the temperature required to convert it to steam increases. So, by putting the cans of food under pressure, much higher temperatures could be used to kill the microorganisms. The water surrounding and inside the food could be heated higher than 212°F without converting to steam.

Refrigeration

People have always known that foods do not spoil as rapidly under cold conditions as in warm temperatures. This is why animals were usually slaughtered in the winter months when the outside temperature was cold enough to prevent spoilage (Figure 12–13).

People also discovered that if foods such as meats were frozen, they could be kept until they were thawed. The first artificial refrigeration made use of ice that was brought down from the high mountains or was saved from winter freezes. If properly insulated, ice could be kept for a long time. Ice caves were dug into the side of a hill, and ice wrapped in insulating materials, such as sawdust, was stored for use in warm weather.

The invention of **mechanical refrigeration** again revolutionized the food preservation industry. Cooled rooms could keep meats, fruits, and vegetables fresh for a longer

© tarheel 1776. Image from BigStockPhoto.com.

Figure 12–13
People have always known that chilled foods do not spoil as rapidly.

BIOTECH in action

Golden Rice-2 Shines in Nutrition Study

All across America, rice has a loyal following among those who enjoy crispy rice cereal at breakfast, steamed white rice with a favorite entree at lunch, or a classic rice pudding as an evening dessert.

But America's consumption of rice—about 21 pounds per person each year—is substantially less than that of people who live in the world's "rice-eating regions," mainly Asia, most of Latin America, and much of Africa.

Because vitamin A deficiency—and its harmful impacts on health—is common in some of these overseas areas, scientists in Europe and the United States have worked for more than a decade to genetically engineer white rice so that it will provide beta-carotene. Our bodies convert beta-carotene into retinol, a form of vitamin A.

White rice typically does not have any detectable beta-carotene. But the genetically engineered Golden Rice-2 from Syngenta Corporation does. Until now, however, scientists have not known how efficiently our bodies can convert the beta-carotene in Golden Rice-2 into retinol.

Research published in a 2009 issue of the *American Journal of Clinical Nutrition* provides a scientifically sound answer. Agricultural Research Service plant physiologist Michael A. Grusak, carotenoids researcher Guangwen Tang, and colleagues reported,

Golden Rice-2 plants growing in a greenhouse.

for the first time, their findings that one 8-oz. cup of cooked Golden Rice-2 provides about 450 micrograms of retinol. That is 50–60 percent of the adult Recommended Dietary Allowance of vitamin A.

Tang, who led the study, is at the ARS Jean Mayer USDA Human Nutrition Research Center on Aging at Tufts University, Boston, Massachusetts; Grusak is with the ARS Children's Nutrition Research Center at Baylor College of Medicine in Houston, Texas.

ARS, the National Institutes of Health, and the U.S. Agency for International Development funded the research.

The scientists based their determinations on tests with five healthy adult volunteers who ate one serving of the rice at the start of the 36-day study. Volunteers' blood was sampled at more than 30 intervals during the research. By analyzing those samples, the researchers were able to determine the amount of beta-carotene (and retinol) that the volunteers absorbed (and then converted to retinol) from the Golden Rice-2.

The efficient conversion of Golden Rice-2 beta-carotene into vitamin A strongly suggests that, with further testing, this special rice might help reduce the incidence of preventable night blindness and other effects of vitamin A deficiency in rice-eating regions. Right now, more than 200 million people around the globe do not get enough vitamin A.

Grusak conducted experiments that made it possible for Tang's group to detect beta-carotene (and resultant retinol) derived from Golden Rice-2, differentiating it from beta-carotene or retinol from other sources.

In his experiments, Grusak determined how to get Golden Rice-2 plants, grown in his rooftop greenhouse at Houston, to take up a harmless tracer and incorporate it into the beta-carotene in the developing grains. The tracer, a rare yet safe and natural form of hydrogen, can be detected by a gas chromatograph-mass spectrometer, the kind of instrument that Tang's team in Boston used to analyze volunteers' blood samples.

The tracer, deuterium oxide, is not new to vitamin A research. But Grusak's studies are the first to show how the tracer can be successfully incorporated into the grains of a living plant for vitamin A investigations.

"It was tricky to determine how much tracer to use and when to add it to the nutrient solution we grew the plants in," says Grusak. His method might be used in other pioneering research geared to boosting the nutritional value of other grains worldwide.

—By Marcia Wood, Agricultural Research Service information staff.

time. When mechanical refrigeration was developed that would freeze foods, people could have access to foods year-round. The cold temperature simply slowed down or completely stopped the actions of any microorganisms in the foods. The invention of the refrigerated rail car and later the refrigerated truck and aircraft meant that fruits, vegetables, and meats could be transported over large distances and arrive fresh at almost any destination in the world. This is why we can enjoy fresh fruits and vegetables all year-round (Figure 12–14). In the winter, produce grown in the Southern Hemisphere can be shipped to us in fresh form.

A significant advancement in the use of frozen food technology was the invention of the **freeze-drying** process. This process makes use of the same laws of physics used to heat cans to a high temperature, except in reverse. Remember that water converts from a liquid to steam at 212°F (100°C) at atmospheric pressure. In canning, pressure is applied to raise the temperature at which water boils; in the freeze-drying process, pressure is removed to decrease the temperature at which water boils. Food products are quickly frozen and then are placed in an airtight chamber. A tight vacuum is pulled within the chamber until the ice in the frozen food begins to change to vapor. In this process, water actually boils at temperatures below 0°F, and eventually all the moisture within the product is removed (Figure 12–15).

Figure 12–14

Because of refrigerated transportation, we can enjoy fresh fruits and vegetables year-round.

Figure 12–15
Freeze-dried foods are frozen and placed in an airtight vacuum chamber. A tight vacuum is pulled in the chamber. These strawberries are freeze-dried.

The freeze-dried product is then sealed in an airtight package and can be stored at room temperature because there is not enough moisture for the growth of microorganisms. When the food is prepared for eating, it is either eaten dry or reconstituted by adding water. Freeze-dried foods are popular with hikers and campers because the food is light to carry.

Enzyme Production

Enzymes are protein-based substances that cause or speed up chemical reactions in living organisms. All living things depend on enzymes to create processes that allow them to live, function, and reproduce. Without enzymes, most of the necessary chemical reactions would not take place within our bodies. For example, enzymes exist in the saliva created in our mouths and in the gastric and pancreatic acids produced in our digestive systems. Enzymes are what make the foods we eat usable by our bodies. Among other functions, they change the sugars, starches, proteins, and other nutrients into a form that can be absorbed by the body and converted to be used as energy or cell replacement. Other enzymes cause the blood to start to clot after a wound has occurred. Others help the body begin to recover after an injury or illness, while others may aid in the reproductive process. In fact, over 3,000 different enzymes have been identified that serve some purpose in the human body.

Humans have made use of enzymes for thousands of years to prepare and preserve foods. Remember from a previous chapter that according to legend, cheese making was discovered in the Middle East when a man stored milk in a pouch made from the stomach of a calf. The pouch was hung over the saddle of a camel and taken on a journey. With the warm temperatures and the sloshing action of the liquid, cheese was formed. The coagulation of the milk curds was started by a naturally occurring enzyme in the calf's stomach. This enzyme, called **rennin**, is what causes milk to coagulate in the stomach of a young calf. It functions by speeding up the process of separating the solids (called **curds**) from the liquid (called **whey**) in the milk (Figure 12–16).

For at least 3,000 years, rennin has been harvested from the stomachs of slaughtered calves and used to create cheese from milk. As scientific knowledge began to grow,

Figure 12–16
Enzymes are used in cheese making to accelerate the process of separating the whey and curds.

the enzymes in rennin were identified and the most important, **chymosin**, was isolated. Then, around 1990, a process of producing this enzyme was developed. Scientists located the exact gene in calves that is responsible for the creation of chymosin. They then transferred the gene into the genomes of bacteria and yeast, which were cultivated in huge quantities, a process called **biomanufacturing**. During biomanufacturing, genetically engineered microorganisms are grown in large tanks with growth conditions (nutrient media, air, pH, and temperature) tightly controlled. The microorganisms produce enzymes and excrete them into the surrounding media, which then may be collected and the enzymes purified. Most of the cheese produced in the United States is made using enzymes from the genetically engineered microorganisms.

Enzymes are used in many other areas of the food industry. Flavors can be enhanced or even created by using enzymes to initiate chemical reactions in the food as it is processed. Lower-quality cuts of meats can be made more tender and tastier through their use. Bread can be stored longer by using an enzyme that slows the chemical reactions that make the bread go stale (Figure 12–17).

The production of enzymes is a worldwide industry bringing in almost $11.2 billion each year. Prior to the use of bioengineering, the cost of producing these materials was quite high. The use of genetically altered microorganisms has

Figure 12–17
Enzymes are used in bread making to help keep the bread fresh.

TABLE 12–1 Enzymes

Enzyme name	Genetically engineered organism	Use (examples)
Alpha-acetolactate decarboxylase	Bacteria	Removes bitter substances from beer
Alpha-amylase	Bacteria	Converts starch to simple sugars
Catalase	Fungi	Reduces food deterioration, particularly egg-based products
Chymosin	Bacteria or fungi	Clots milk protein to make cheese
Cyclodextrin-glucosyl transferase	Bacteria	Modifies starch and sugar
Beta-glucanase	Bacteria	Improves beer filtration
Glucose isomerase	Bacteria	Converts glucose sugar to fructose sugar
Glucose oxidase	Fungi	Reduces food deterioration, particularly egg-based products
Lipase	Fungi	Modifies oil and fat
Maltogenic amylase	Bacteria	Slows staling of breads
Pectinesterase	Fungi	Improves fruit juice clarity
Protease	Bacteria	Improves bread dough structure
Pullulanase	Bacteria	Converts starch to simple sugars
Xylanase (hemicellulase)	Bacteria or fungi	Enhances rising of bread dough

Source: Cornell Cooperative Extension Service.

greatly benefited the food processing and preservation industry. Table 12–1 lists the enzymes that are currently manufactured using genetically engineered bacteria and fungi.

Summary

Since the very beginning, our story of human existence has been that of the struggle to find enough food to eat. In modern times, a large industry has developed that delivers food for all of us to eat. Biotech scientists continue to make advances in developing foods that are more nutritious, higher yielding, and easier to grow and ship. Food science and technology has given the United States the most bountiful, safe supply of food that humans have ever known. However, millions of people living in developing countries do not have enough food and rely on staple crops that lack important macro- and micronutrients. Golden Rice and BioCassava Plus are examples of biofortified crops in the pipeline that will be released to global markets in the near future, helping to alleviate malnutrition and hunger. Early forms of biotechnology, such as the use of salt, yeasts, and bacteria and other microorganisms, have been used to prepare and preserve foods for centuries. With the use of enzymes produced through biomanufacturing, biotechnology continues to play a vital role in food production and preservation.

CHAPTER REVIEW

Student Learning Activities ....

1. Select a developing country where dietary deficiencies are a real problem. Locate information about the country and its problems. Try to determine the cause of the dietary deficiencies within that particular country. What is the common diet of the people? What nutrients are missing? Think of ways in which biotechnology can be used to solve the problem. Share your report with the class.

2. Conduct further research on Golden Rice. Locate articles written that are critical of the concept. Organize your thoughts on how you feel about Golden Rice. Conduct a debate in class between proponents and opponents of the concept.

3. From Table 12–1, select a commercially produced enzyme and report to the class on how it is produced and used. What are the benefits of using the enzymes produced by genetically modified organisms over the enzymes produced by conventional means?

Fill in the Blanks ....

1. Our bodies need a wide variety of different _____ to perform such functions as cell _____, the development and maintenance of our _____ systems, and the generation of the _____ we need in our daily lives.

2. _____ are plant-derived molecules that may have _____, _____, or _____ enhancing properties.

3. A shortage of either _____ or _____ can cause our bodies to _____, which in turn leads to _____ or even death.

4. _____ _____ may be made in the human body by metabolizing the phytochemical _____, which provides the _____ or _____ color to fruits and vegetables.

5. In addition to _____, Golden Rice researchers are now working on increasing levels of _____, _____, _____ and _____ in this biotech crop.

6. Early forms of _____ played a crucial role in the _____ of food crops and animals by providing the _____ to properly preserve food until it can reach our tables for _____.

7. The process of _____ uses heat to kill _____, while the process of _____ removes the moisture necessary for bacterial growth from the food.

8. The invention of the _____ rail car and later the refrigerated truck and _____ meant that fruits, vegetables, and meats could be _____ over large distances and arrive _____ at almost any destination in the world.

9. _____ are protein-based substances that _____ or speed up _____ _____ in living organisms.

10. Biotech scientists continue to make advances in developing foods that are more _____, higher _____, and easier to _____ and _____.

True or False

1. To meet the needs of a growing world population, we will need to increase food production by 25 percent by the year 2050.

2. A biofortified crop is a crop that has been genetically engineered to contain higher-than-normal levels of nutrients needed by humans.

3. Nutrient deficiencies are a major problem in developed countries because people in developed countries often cannot afford diets high in fresh fruits and vegetables.

4. Rice and cassava are two important staple crops for people in developing countries.

5. The developers of Golden Rice are expecting to make a huge profit from the sales of the crop.

6. People have known for centuries what causes food to spoil.

7. Canning was developed by the French during the nineteenth century.

8. Enzymes are what makes the food we eat usable to our bodies.

9. Rennin is the most important enzyme in chymosin.

10. The use of genetically altered microorganisms has made food processing and preservation much more affordable.

Discussion

1. Identify the difference between macronutrients and micronutrients. Give an example of each type of nutrient.

2. Explain the functions of proteins, carbohydrates, and lipids in the human diet.

3. What problems do dietary deficiencies of vitamin A and iron cause in humans?

4. What is Golden Rice?

5. Why was Golden Rice developed?

6. Explain why drying and salting preserve food.

7. What were two major problems with the original process of canning food?

8. Explain how the invention of mechanical refrigeration revolutionized the food industry.

9. Describe the process of freeze-drying food.

10. Describe at least two ways that enzymes are used in the food industry.

CHAPTER 13

Biotechnology in Ecology

OBJECTIVES

When you have finished studying this chapter, you should be able to:

- Define ecology.
- Discuss the role biotechnology plays in benefiting our environment.
- Explain how biotechnology is used to detect environmental pollutants.
- Distinguish between bioremediation and phytoremediation.
- Discuss high-yield farming and its benefits to the environment and world hunger.
- Explain how genetically modified crops positively affect the environment.
- Discuss conventional farming versus organic farming and the advantages and disadvantages of both in relation to the environment.
- Explain how oil spills can be combated with biotechnology.
- Discuss how biodiesel and biofuels benefit the environment.
- Discuss the benefits of growing plants that can survive harsh conditions.

Figure 13–1
Humans have the responsibility of caring for the ecology of the earth.

Ecology

Throughout this text, you have read of the marvels of modern biotechnology and the problems solved by the new discoveries and developments. The impact of biotechnology in the area of **ecology** is rapidly increasing. Ecology is defined as a branch of science concerned with the interrelationship of organisms and their environments. More simply, it is how all organisms interact within our home—our environment on this fascinating, complex spaceship we call Earth. As inhabitants of the earth, it is our responsibility to care for the environment and all the interactions of the organisms that live here (Figure 13–1).

Because of the growing global pressures on air, water, and land resources in recent years, environmental concerns have received a lot of attention. These pressures are the result of an ever-increasing population that must eat, breathe, consume water, and dispose of waste material. Since humans produce so much waste material, contamination of the environment is a concern for all who live in the environment, and steps must be taken to cleanup any pollution. Cleaning up the environment is a very expensive, time-consuming task, and scientists are constantly looking for less expensive, more efficient ways to accomplish the removal of contaminants (Figure 13–2).

In order to reduce some of the costs of research and development and to make cleanup more efficient, new biological techniques are being developed that will have a positive

Figure 13–2
People produce a lot of waste. If we do not constantly clean it up, the environment will become contaminated.

impact on our environment. In this chapter, the problems facing our environment and the advancements that have been made in biotechnology to assist in the elimination of these problems are discussed.

Agriculture and Our Environment

Agriculture is an important tool in the sustainability of our environment. The plants we grow, the chemicals we spray, and the equipment we use can all have an important impact on the safety of the air we breathe and the water we drink. Since we have to live in the environment and we have to have agriculture to provide the food, shelter, and clothing we need, it is vitally important that we find ways that agriculture can help the environment and cause little harm (Figure 13–3).

Advances in biotechnology within agriculture can greatly improve our environment. For example, in order to decrease the amount of pollution in our waters, agricultural scientists and researchers have determined ways that plants and bacteria in the soil can absorb toxic wastes. Oil-dissolving bacteria, used to combat oil spills on beaches, are one of the latest discoveries in cleaning up our oceans and protecting our marine life. As you remember from Chapter 9, significant advancements have also been made in gaining more crop yield per acre in order to decrease the amount of deforestation and feed the growing numbers of Earth's inhabitants. Crops have also been genetically modified to resist

© iStockphoto/DariuszPA

Figure 13–3
Agriculture and the environment are two areas of science that must work together.

harmful diseases and insects, reducing the need to spray large quantities of pesticides and fungicides that linger in soil and pollute waterways.

Detecting Environmental Pollutants

Biotechnology is playing a large part in detecting or monitoring pollution and in determining how much is present. Biological systems such as living organisms, animal immune systems, and enzymes are being used to detect pollution.

Using **indicator species** is one of the oldest methods of biological detection. This method uses plants, animals, and microbes to warn us about pollutants in the environment. These species are used to determine the impact of various agricultural practices, ranging from chemicals to transgenic plants. The impact may be determined by the presence or absence of certain species (Figure 13–4). For example, the absence of certain insects from a stream may be an indicator that the stream is polluted. In fact, when any plant, animal, or microbe that normally grows in an area begins to disappear, suspicions are raised that pollution has occurred.

Immunoassays are tests that use antibodies from animal immune systems to detect specific pollution compounds. They can detect pollutants in amounts as small as a few parts per billion. Immunoassay kits are commercially available to detect many common agricultural chemicals that have

Figure 13–4

The absence of certain species may indicate that an area is polluted. The presence of frogs may indicate a clean wetland area.

seeped into our soil and water. The kit contains many types of antibodies derived from animals that can detect the presence of certain chemicals by changing color. If the antibodies in a compound change to a certain color, the chemical or compound is present in the soil or water being tested.

Biosensors are devices that combine biological and electronic systems to detect or measure small amounts of specific substances, including environmental pollutants. They have been widely used in medicine and are now being used in environmental situations (Figure 13–5). They may be applied to a wide variety of contaminants and situations, including *Escherichia coli* bacteria on food, chemicals in soil, and biological oxygen demand in water. The biosensor's biological component always produces some type of physical change in response to the environment. An electronic instrument called a **transducer** measures that change. Biosensors can be designed to be very selective or broad spectrum. Selective biosensors are sensitive to just one compound, whereas broad-spectrum biosensors are sensitive to a wide range of compounds. For example, a broad-spectrum biosensor measures changes in algal chlorophyll in stream water. The changes reflect stresses on the plants that may come from different pollutants or a combination of them. Another example is using biosensors to detect trace amounts of herbicides that inhibit photosynthesis in plants.

Figure 13–5
This scientist is using a biosensor to measure toxins in corn.

Bioremediation

An important area of biotechnology is **bioremediation**. This is defined as a set of techniques that use living organisms to cleanup toxic wastes in water and soil. The word is derived from the Latin words *bios*, meaning "life," and *remedium*, meaning "fix" or "cure." In nature, the environment is constantly being cleaned. When a plant or animal dies, microbes begin to decompose the plant or animal as a way of disposing of the dead tissues (Figure 13–6).

When microbes (organisms found everywhere in nature) eat toxic wastes, they turn them into harmless compounds, such as carbon dioxide and water. They die for lack of food or return to their original levels after they finish transforming the wastes. In some cases, scientists add nutrients, such as nitrogen and phosphorus, to stimulate the growth of

Figure 13–6
When an organism dies, microbes begin to break the tissues down. This is nature's way of cleaning the environment.

Figure 13–7
Contaminated soil is sometimes excavated and taken to a treatment site. This is an expensive operation.

these naturally occurring microbes. This technique is called **biostimulation** and is used to rapidly increase the amount of beneficial microbes.

Most often, polluted soil is treated at its "on-site" location with bioremediating microbes but can also be excavated and taken to a treatment site (Figure 13–7). Microbes may be used to cleanup a variety of pollutants, including insecticides, fungicides, herbicides, petroleum products, and some detergents. Researchers have developed ways to determine which microbes are the best at destroying certain toxic wastes by first looking at those that are found at the pollution site. Sometimes, a mixture of microbes is found to be the most effective recipe for treating certain wastes.

A method has been developed that may prove to decrease the amount of toxic waste at a more accelerated rate. **Enhanced bioremediation** is the term given to the group of techniques in which nutrients, microorganisms, or other materials are introduced to a contamination site to accelerate the cleanup process.

In order to determine if bioremediation is the most efficient way to decrease the level of toxic waste in soils, several tests are conducted both on-site and in laboratories. Samples of the material are treated with the active microbes and therefore allow researchers to design the best approach for the specific site. After the soil is treated, the site is studied periodically to determine if there is a need to retreat. When levels of the contaminant have reached target goals, state and regulatory agencies study the site and consider it for closure.

Bacteria are among the nation's chief recyclers. Because of their ability to break down a variety of compounds into their basic elements, bacteria are used extensively in environmental biotechnology. One of the applications where bacteria are gaining greater use is in the oil industry. Every day, there are billions of gallons of crude oil transported across the world's oceans in huge tanker ships. When these ships meet with an accident, a tremendous amount of oil is leaked into the water and is often washed ashore (Figure 13–8). Here,

© Malcolm Fife/Photodisc/Getty Images

Figure 13–8
Every day, billions of gallons of crude oil are transported by ships. Sometimes, accidents occur.

Figure 13–9
Numerous species of oil-degrading bacteria were discovered at the Rancho La Brea tar pits in California.

the oil pollutes the environment, causing severe damage to plant and animal life. Cleaning up an oil spill is a complicated, time-consuming, and expensive operation.

Biotechnology offers great promise in the development of oil-consuming bacteria. There are more about 9,000 known, culturable species of bacteria that inhabit our environment, and of these, over 1,300 now have complete genome sequences. However, microbiologists have estimated that there are millions of unculturable species that we have yet to characterize. When scientists are looking for microbes with special characteristics, such as the ability to break down oil, they look in natural environments where the bacteria are present. In 2007, for example, numerous species of oil-degrading bacteria were discovered at the Rancho La Brea tar pits in California (Figure 13–9).

Several different types are capable of breaking down both simple and complex hydrocarbons (organic compounds that contain only hydrogen and carbon), the components of crude oil. These bacteria are called **oleophilic**, meaning that they are attracted to oil and can be found in areas that naturally contain oil or where oil is present. The bacteria simply break down the hydrocarbons found in oil that have no obvious use. The bacteria then convert the hydrocarbons into methanol (a type of alcohol), water, and carbon dioxide.

Oleophilic bacteria can both consume and dissolve the heavy sticky oil known as chocolate mousse that clogs up beaches after a shipwreck or oil spill (Figure 13–10). The

Figure 13–10
After an oil spill, oleophilic bacteria can consume and dissolve the heavy oil that coats the shore.

bacteria eat up some of the oil and turn the remainder into a less viscous "oil milk" of tiny oil droplets in the water. The more finely divided the oil is, the more accessible it becomes to other types of oil-eating bacteria. The great advantage of using bacteria to combat oil slicks is that we do not have to use chemicals or surfactants, such as detergents, to dissolve the oil. These substances are expensive and may have a detrimental effect on the environment.

Scientists and researchers across the world are testing other items that may be useful in cleaning up oil spills. One such item is tree bark. The bark is considered an absorbent and makes it easier to lift the oil by means of spades and other tools. A further effect is that by covering the spilled oil, the bark prevents animals from coming into contact with it.

By incorporating these aspects of biotechnology to combat oil spills, we are posing less threat to humans, beaches, the water supply, and aquaculture. People can enjoy their visits to the beaches, the beaches can remain a safe home for their natural inhabitants, our water supply is less likely to become contaminated, and sea life can grow and thrive with a smaller threat of oil spills damaging their habitat.

Phytoremediation

Phytoremediation is the process of plants or trees absorbing or immobilizing pollutants. The prefix *phyto* means "plant." Trees and plants are being used to cleanup soils by absorbing more complex materials than can be consumed by soil microbes (Figure 13–11). Such materials include heavy metals, solvents, hydrocarbons, pesticides, radioactive metals, explosives, nitrates, crude oil, runoff from landfills, and organic pollution that could burden surface waters.

Plants are used primarily where contamination is slight to moderate and does not extend much deeper than the roots of the plants. Once the soil and water have been cleaned, the plants can be removed and replaced with fresh plants. The process of phytoremediation can take as little as weeks and as long as years to achieve cleanup standards. They can also be used in conjunction with bioremediation and its microbial techniques.

© Courtesy of ARS/USDA Image #K9054-9. Photo by Keith Weller

Figure 13–11
Plants can sometimes be used to cleanup pollution. This plant is an alpine pennycress that removes excess zinc and cadmium from soils.

Advantages and Limitations of Using Bioremediation and Phytoremediation

There are many economic and environmental advantages as well as limitations to using these cleanup processes. Advantages include low cost, the ability to harness natural processes, reduction of environmental stress, the use of attractive plants, and early use.

First, the low cost of these processes is attractive to environmentalists. Most traditional technologies require the removal and disposal of contaminated soils. However, bioremediation usually can be carried out in place, resulting in a much lower cost. This process has been found to cost anywhere from one-fifth to one-half as much as other nonbiological techniques. Second, natural processes can be used to cleanup without human intervention. The only costs normally associated with these cases are testing and periodically monitoring the site. Another advantage of bioremediation and phytoremediation is that they reduce environmental stress. The processes tend to minimize site disturbance, preserve topsoil, and reduce postcleanup costs.

Phytoremediation has several advantages because the use of plants can be visually appealing and involves solar energy to carry out pollution degradation. This process can also be used as soon as a contaminant is identified, while other forms of site identification and testing can last up to 3 years.

Although there are advantages to using the techniques, there are also limitations to the processes of bioremediation and phytoremediation. These drawbacks include time, inapplicability to certain contamination situations, and public concerns.

Bioremediation techniques, in particular, are considered to be significantly slower than chemical techniques. However, as more modern advances are made with these studies, better use of natural processes and newly engineered organisms will speed up the processes (Figure 13–12).

The second drawback to these new techniques is that they are not applicable in all situations. For example, bioremediation is used on only a small percentage of all sites undergoing cleanup. As more knowledge is gained on how microbes work within our soil, more applications will be discovered. Finally, many people feel as though bioremediation is not the right method to use when decreasing toxins within our soils. Often, those who live near treatment

Figure 13–12
Bioremediation techniques are currently slower than chemical techniques, but research promises to speed up the processes.

sites would much rather have the contaminated soil removed than treated on-site. The fear is that either the process will not decontaminate the soil or the organisms used will have a detrimental effect on the soil and environment.

High-Yield Farming

Modern **high-yield farming**, which began as the Green Revolution in the 1960s, was a significant environmental triumph. During the 1960s and 1970s, it not only saved hundreds of lives but also spared millions of square miles of wilderness from conversion to farmland (Figure 13–13).

Figure 13–13
The Green Revolution prevented millions of square miles of land from being cleared for farming by increasing yields on land already in production.

The solution to preventing global starvation and wildland destruction in the first half of this new century is to revitalize the Green Revolution through high-yield farming—producing more per acre.

High-yield practices are also applied to forestry. High-yield forestry meets human demands for forest products with significantly lower land requirements, allowing for wider conservation of natural forests and the array of wildlife and plant species within those forests.

The world's population will likely rise to 9 billion people before 2050. Worldwide per capita consumption of meat, dairy products, fruits, and vegetables is increasing rapidly as living standards rise throughout the world. One of the greatest threats to the earth's biodiversity is habitat loss through the conversion of natural ecosystems to agriculture (Figure 13–14).

There are only two ways of increasing food production in order to feed the world: either grow more on the land now in cultivation or clear new land. Without increasing crop yields from available land, wildlands and species will be destroyed in order to keep human populations from starving. Some 2 billion of the earth's poor live in or near forests that are home to three-fourths of the world's wildlife species. Without a new Green Revolution, the only way they can feed their families is to burn down more forests to cultivate farmland. In addition, global demand for forest products is increasing rapidly and may double over the next half century. Biotechnology offers a way to help solve the problem.

Figure 13–14
The conversion of natural ecosystems to agriculture poses great danger to our world's biodiversity.

© Steve Collendar/www.Shutterstock.com

Figure 13–15
One of the long-term threats to human existence is soil erosion.

One of the long-term threats to human existence is soil erosion (Figure 13–15). All life on Earth is dependent on the soil to grow plants that sustain animal life by providing food and replenishing the air with oxygen. Doubling the yields on the best and safest farmland cuts soil erosion by more than half. Herbicide and conservation tillage can cut those low rates of soil erosion by 65–98 percent. In addition, high-yield producers are in the midst of developing "no-leach" farming. Tractors and applicator trucks for farm chemicals now can be guided by global positioning satellites and radar within inches of their true positions across the field while varying the application rates of chemicals and seed based on soil sampling, soil hydrology, slope, plant population, and nearness to waterways. It is now practical to manage our farms by the square yard rather than in chunks of 10 or 100 acres.

The world's severe soil erosion today is occurring in primitive countries that are trying to support rising populations by extending low-yield farming into fragile lands. Genetic modification of plants, such as breeding plants to be insect resistant and herbicide tolerant, has two main advantages: producing biodegradable industrial raw materials and helping to cleanup the environment.

Since biotechnology has developed these practices, the amount of chemicals required by farmers for pest and weed control has dramatically decreased. For example, a single, less toxic herbicide can be used in the place of several herbicides,

Helping Earth-Friendly, Corn-Based Plastics *Take the Heat*

Those little plastic coffee-cup lids that keep your steaming hot java from sloshing all over you might someday be made from biodegradable, corn-based plastic. Making corn-derived plastics more heat-tolerant, so that they would not distort when they are not supposed to, is one of several top-priority targets of collaborative research under way since 2007 at the Agricultural Research Service's Western Regional Research Center in Albany, California, near San Francisco.

Chemist William J. Orts, who leads the center's Bioproduct Chemistry and Engineering Research Unit, works with collaborators Allison Flynn and Lennard F. Torres of Santa Barbara-based Lapol, LLC, to broaden the range of applications for which corn-based plastics would be ideal. The partnership, carried out under terms of a cooperative research-and-development agreement, aims at making those plastics an alluring alternative to petroleum-derived plastic goods.

Plastics made from corn are generally biodegradable, and corn is a renewable natural resource. While plastics made from petrochemicals can be biodegradable, they are derived from a finite, nonrenewable source.

What is more, manufacture of corn plastics usually causes less pollution, including fewer greenhouse gas emissions, than does the production of petro plastics.

Corn-based plastics are made by fermenting corn sugar to produce lactic acid. In turn, the lactic acid is used to form a bioplastic known as "polylactic acid," or PLA. Trouble is, PLA has a lower heat tolerance than some petroleum-based plastics. "That excludes PLA from being used for some applications," says Flynn.

In the plastics industry, the upper limit of PLA's heat tolerance, that is, the temperature at which it may begin to distort, is referred to as its "heat-deflection temperature."

To overcome this obstacle, Flynn, Torres, and Orts are developing a new product. Known as a

ARS chemist William Orts (left) works with collaborators Allison Flynn and Lennard Torres of Lapol, LLC, to improve the heat tolerance of environmentally friendly plastic made from corn.

"heat deflection temperature modifier," it can be blended with PLA to make PLA more heat tolerant. The modifier itself is more than 90 percent corn based and is fully biodegradable, according to Flynn.

Preliminary tests at the Albany lab indicate that, when blended with PLA, the modifier can raise PLA's heat-deflection temperature by at least 50°F. With further research and development, the heat modifier might make it possible for hundreds of products that currently can't typically be made with PLA to one day be manufactured with this bioplastic.

The products of the future might include not only coffee-cup lids, but perhaps food or beverage bottles or other containers that are "hot-filled," that is, filled at the food-manufacturing or beverage-bottling plant while the food or beverage is still hot from pasteurization. Examples might include, among other popular items, tomato catsup or some kinds of fruit juice. Today's PLA typically cannot take the heat of hot-filling.

ARS and Lapol are currently seeking a patent for the invention.

"Right now, there are no commercially available heat-deflection temperature modifiers for PLA," says Lapol chief operating officer Randall A. Smith. "It's an emerging market."

The encouraging preliminary results for the experimental heat-deflection temperature modifier suggest that new opportunities for PLA may indeed be heating up.

—By Marcia Wood, Agricultural Research Service information staff therefore providing more efficient weed control and potentially lower cost. A new method of farm management, called **integrated farm management**, aims to reduce applications of chemicals by optimizing the combination and timing of all farm management activities.

Biotechnology techniques involve the use of diagnostic kits that provide accurate, sensitive, quick, and reliable diagnosis of certain diseases. Producers will be able to detect specific pests more easily and apply the appropriate quantity and correct pesticide to combat plant disease at an earlier stage. Agrochemicals can then be applied only when required, potentially reducing the total amount of chemicals applied to the land.

Herbicide-resistant crops also help the environment by reducing the amount of cultivation needed. When a field is plowed, the turning over and stirring of the soil spurs the release of carbon dioxide, which has been identified as a greenhouse gas contributing to global warming and climate change. When a field is not tilled, the decomposition and release of carbon dioxide is vastly slowed. Plants trap carbon dioxide during photosynthesis, and that carbon goes into the soil after harvesting. The object of no-till farming is to keep carbon in the soil rather than release it into the atmosphere as carbon dioxide. Also, the organic matter left sitting on the soil works effectively to decrease water runoff and erosion and boost the soil's nutrient retention (Figure 13–16).

Figure 13–16

Conservation or no-till farming can help prevent the erosion of the soil.

Another environmental benefit of no-till farming is the reduction of fossil fuels being emitted into the atmosphere by agricultural machinery. Tractors are used less frequently, which in turn increases our air quality.

According to a U.S. Department of Agriculture survey taken in 2009, about 88 million acres in the United States were no-till, or a total of about 35.5 percent of the country's farmland. The percentage of no-till acres is increasing by about 1.5 percent each year when looking at the eight major crops cultivated in the United States. Adoption of no-till farming is helping to significantly reduce U.S. carbon dioxide emissions. In addition to environmental benefits, there are also benefits to farmers. Labor and time are saved when the farmer does not plow the field, and productivity is increased.

Environmental Concerns of Genetically Modified Crops

Whenever new technology is introduced, there are concerns about its safety and effectiveness. Environmental concerns about the use of biotechnology in agriculture focus on the transfer of "new" genetic material to wild plants or animals. Specifically, the concerns are that herbicide-tolerant crops will help create resistant strains of weeds or that insect-resistant potatoes will result in insects tolerant to the natural insecticide.

Genetic material can be transferred from domestic to wild plants. This happens with both biotech plants and those developed through conventional plant breeding. Scientists are considering certain strategies that may help reduce the possibility of resistant weeds and pests. One such strategy is to plant genetically modified (GM) crops in areas where there are not closely related wild species or cultivated domestic species nearby (Figure 13–17). Another strategy is to develop an alternative method of pest control in case a population of resistant weeds or insects develops. A third strategy is to plant what are called **refugia**. This is a process in which non–insect-resistant plants are grown nearby, either mixed with the biotech crops or planted in large sections. A population of nontolerant insects is then available to mate with those

Figure 13–17
Planting crops that are of different species helps reduce the possibility of genetic transfer.

that might develop resistance. Researchers are also working toward having the natural insecticide become present in the plant only when the plant is at its most susceptible. The release of the insecticide in the plant would be triggered at an appropriate time in the plant's growth cycle. In regard to herbicides, farmers are encouraged to cultivate a buffer zone around the herbicide-tolerant field so that the crop is less likely to mix with weeds at the edge of the field.

Growing Crops in Harsh Conditions

Many major agricultural research companies are pouring resources into developing plants that can withstand drier, colder, or saltier conditions. For instance, scientists are working on a breed of rice that can survive long periods under water, a tomato plant that can grow in salty soils 50 times higher than conventional plants can withstand, and corn that can tolerate aluminum in the soil (Figure 13–18). The development of plants that can grow in tough conditions makes it easier to farm marginal lands, helping to keep fragile soils such as wetlands and rain forests out of food production. Using this process also cuts down on the amount of irrigation needed to water crops.

One of the key discoveries has been that most plants have some ability, controlled by their genes, to tolerate cold and dry conditions. Microbiologists have found that by turning

© C. Borland//Photolink/Photodisc/Getty Images

Figure 13–18
Some plants are able to withstand harsh climates. This ability may be genetically transferred into crops.

up the activity of several of the genes or moving genes from hardier organisms, they can make the plant adapt to increasingly hostile conditions.

Biodiesel

Biodiesel, obtained from renewable raw materials, is a non-polluting, biodegradable liquid fuel that can be used to replace fossil diesel fuel in our agricultural machinery. It is made from oilseeds such as soybean and canola oil and has been proved to decrease harmful emissions while enabling the growth of local, environmentally friendly energy supplies. Through the use of agricultural crops used to produce biodiesel in the United States, our dependency on imported oil will drastically decrease (Figure 13–19). If the United States uses just 10 percent of its cropland for the production of biodiesel, all the fuel needed for our agricultural machinery would be produced domestically.

Biodiesel, used as a liquid fuel, offers many environmental advantages over current usage of fossil diesel. They include reducing polluting greenhouse gas emissions, reducing global warming by decreasing carbon dioxide levels in the atmosphere, and obtaining a biodegradable product that degrades into organic material by 85 percent in 28 days. There are many different raw materials that can be used for obtaining biodiesel, such as vegetable oils, used cooking oils, and animal fats.

Figure 13–19
These scientists are developing a less expensive way to make biodiesel fuel.

However, there is one major challenge facing large-scale production and commercial use of biodiesel: it costs three times as much to produce as petroleum diesel. Biotech research is under way to increase the oil content in canola and soybean crops that could make biodiesel production more economically efficient.

Summary

The relationships among agriculture, biotechnology, and the environment are getting closer each day. Natural resources on our planet are diminishing at a rapid rate. However, biotechnology in our agricultural sector may be the answer to our ever-increasing environmental concerns. It is an excellent way to incorporate science with nature to achieve the best possible result—cleaning up our environment. In the next few years, much progress can be made in solving the problem of world hunger. Also, deforestation and destruction of wildlife habitat could be dramatically curtailed through the use of modern biotechnology.

CHAPTER REVIEW

Student Learning Activities

1. Search the Internet for an example of how biotechnology is helping to cleanup the environment. Report to the class on your findings.

2. Choose an underdeveloped country and research the agricultural practices there. Are their practices harmful to the environment? What could be done differently to benefit the population as well as the environment? Are they faced with a hunger problem? Using the practices discussed in the chapter, which would be most beneficial to solving the hunger issue?

3. Locate information on a recent oil spill. Determine the negative effects that the spill had on the environment. Find out if oleophilic bacteria were used in the cleanup.

4. Locate an area in your community that is polluted. Working with your classmates, devise a plan for cleaning up the problem.

Fill in the Blanks

1. _____ is defined as a branch of science concerned with the _____ of _____ and their _____.

2. _____ are tests that use _____ from animal _____ systems to detect specific _____ compounds.

3. _____ is defined as a set of techniques that use _____ _____ to cleanup _____ wastes in water and soil.

4. The great advantage of using _____ to combat oil slicks is that we do not have to use _____ or _____ such as _____ to dissolve the oil.

5. _____ is the process of plants or trees _____ or _____ pollutants.

6. Limitations to the processes of _____ and _____ include _____, _____ to certain contamination situations, and _____ concerns.

7. _____-resistant crops also help the _____ by reducing the amount of _____ needed.

8. The development of plants that can grow in _____ conditions makes it easier to farm _____ lands, helping to keep _____ _____ such as wetlands and rain forests out of food _____.

9. _____, obtained from _____ raw materials, is a _____, _____ liquid fuel that can be used to replace _____ diesel fuel in our agricultural machinery.

10. If the United States uses just _____ percent of its cropland for the production of _____, all the fuel needed for our _____ machinery would be produced _____.

True or False

1. Indicator species are organisms that can be used to warn us about pollution in the environment.

2. Broad-spectrum biosensors are sensitive to just one compound, whereas selective biosensors are sensitive to a wide range of compounds.

3. The technique of biostimulation is used to rapidly increase the number of indicator species in an area.

4. Bacteria are used extensively in environmental biotechnology because of their ability to break compounds down into their component elements.

5. Researchers are testing the effectiveness of tree bark in cleaning up oil spills.

6. Phytoremediation is used where pollutants are extensive and extend deep into the soil.

7. Many people do not want to use bioremediation to treat contaminated soil because they feel that it will have a detrimental effect on the soil and the environment.

8. Increasing yields on the best farmland will also increase soil erosion in those areas.

9. Refugia are non–insect-resistant plants that are grown either nearby the biotech crop or mixed with the biotech crops and are used to reduce the development of pesticide-resistant insects.

10. Biodiesel is far cheaper to produce than traditional petroleum diesel.

Discussion

1. Why is it important for researchers to find ways for agriculture to become more sustainable?

2. What are biosensors, and how do they work?

3. Explain the process of bioremediation.

4. Describe the role oleophilic bacteria play in bioremediation.

5. What are the advantages of bioremediation and phytoremediation?

6. Explain the process of "no-leach" farming and explain why this practice will decrease the environmental effects of agriculture.

7. Describe the use of integrated farm management.

8. How does the use of no-till farming help reduce climate change?

9. How might the use of biotechnology lead to herbicide-resistant weeds?

10. How could the use of biodiesel have a long-term positive effect on the environment and the economy?

CHAPTER 14

Consumer Concerns about Biotechnology

OBJECTIVES

When you have finished studying this chapter, you should be able to:

- Discuss the arguments given against the use of bioengineered foods.
- Describe various concerns with the use of bioengineered livestock.
- Explain the concerns that consumers have about cloning, both animal and human.
- List the regulatory agencies and current safeguards used to monitor the use of bioengineered organisms.

Introduction

Modern biotechnology is a rapidly changing field that is constantly creating new advancements and technologies. Each advancement brings about new concerns, especially in the area of **genetically modified organisms (GMOs)**. Modern consumers try to be informed about what goes into their food and the happenings that affect the environment. Consumers want to know exactly what they are getting in terms of food and what effects any new technology will have on us as humans as well as on the world we live in (Figure 14–1). Some people may fear the unknown and may not be comfortable with a product simply because it is derived from new technology or from methods that we as consumers do not yet fully understand. Others may have misgivings about the effects that biotechnology could have on society and on our surroundings.

Food Issues

The primary consumer concerns within the biotechnology field today revolve around food, environmental concerns, and the current debate surrounding the development of cloning. Many issues have been raised concerning bio-engineered food. Some people visualize foods that result from genetically altered foods as being monstrous. In fact,

Figure 14–1
Consumers want to know exactly what they are getting in their food.

Figure 14–2
Some people refer to food from genetically modified organisms as "frankenfood."

the term "frankenfood" has been coined to describe food from GMOs (Figure 14–2). This is perhaps an outgrowth of science fiction that portrays genetic modification of any organism as having catastrophic results.

We are products of what we ingest, and we naturally have concerns about what we are putting into our bodies and how it will affect not only us but future generations as well. Much debate has surrounded plant biotechnology such as **genetically modified organisms (GMOs)**. Many citizens of the United States and countries such as those in Europe have tried to stop the marketing of GMOs. Concerned people all over the world have been trying to judge the safety of food derived from genetically modified plants. The concern of many is the effect, if any, that products such as GMOs may have on their health and the environment. The dilemma is whether a product that still has what some consider to be unknown outcomes should be produced.

Whether altered genetically to improve the taste or shelf life or to increase crop resistance to insects, most of the food that we eat has been affected by science in some fashion. We have greatly altered nearly every domesticated crop plant species from its original, natural form. Crossbreeding plants has been the traditional method of improving food source quality and longevity, but it is not precise, and it involves a lot of time and trial-and-error methods. In fact, all the food we eat comes from organisms that have at some time or other been genetically altered by crossbreeding. For example, the corn we eat is vastly different from the original maize found in nature (Figure 14–3). The carrots we grow in our gardens or buy at the grocery store were derived from a common weed called Queen Anne's lace. All the agricultural plants and animals we grow are the result of many years of selection and crossbreeding. Proponents of GMOs argue that the use of gene splicing is much safer and more precise than a crossbreeding program because specific genes can be targeted. A crossbreeding program may have many genes altered that we know nothing about, while gene splicing allows scientists to place a specific gene on a chromosome, controlling a specific desired characteristic. This precision allows for more control and safety.

Critics argue that while biotechnology is a more efficient way to get these desired results, it raises concerns about what unnatural things might be added to our food supply, possibly introducing new allergens or having long-term effects on

© Kent Knudson/Photolink/Photodisc/Getty Images [A] and Steve Cole/Photodisc/Getty Images [B]

Figure 14–3
The corn produced by farmers today (A) is very different from the maize that grows wild (B).

© dogjlikehorse/www.Shutterstock.com

Figure 14–4
It is estimated that 1.5 percent of the U.S. population suffers from some type of food allergy.

our bodies, that we have yet to determine. Of particular concern is the creation of transgenetic organisms that may insert an animal gene into a plant or vice versa. Estimations are that 1.5 percent of the U.S. population suffers from some type of **food allergy** (Figure 14–4). A large percentage of allergens are proteins. Many of these allergens have been identified, but many others are less known and not fully understood. Genetic engineering involves the expression of new proteins. If a new protein created through genetic engineering is similar to a known allergen, it can be tested for any allergic reactions that it might cause. However, if an unknown protein is introduced, it can be quite difficult to assess whether it can cause allergic reaction and to what extent.

A good example of the concern is the case in which a trace of genetically modified corn showed up in taco shells sold by a popular fast-food chain. Although the corn had been cleared for use as livestock feed, the **Food and Drug Administration (FDA)** had not cleared it as a food for humans. Although no allergies were known to result from eating the corn, the tacos were pulled off the market, and several lawsuits followed. This action had a tremendous effect on the corn market, as many tons of grain were pulled off the market and thousands of corn products were removed from grocery shelves (Figure 14–5). Advocates of GMOs felt that the action was taken not because of problems caused by the genetically altered corn but by fear of the unknown on the part of consumers.

Figure 14–5
Tacos made from corn with a trace of genetically modified corn were removed from the market because of consumer fears.

It is unlikely that harmful substances or genes for known toxins would be intentionally transferred into an animal destined for consumption, but what about the by-products of genetic engineering or unknown toxins that might be transferred into a food animal? For example, many genetically engineered fish have had a growth hormone gene introduced to speed up the growth process. It is very important for us to ensure that these products have no adverse effect on the humans or animals that consume them.

Many consumers are apprehensive about new substances being introduced into our food supply because of the "unknown" factor. They question whether we know that these new substances are safe for consumption. They also have concerns about the long-term effects genetically altered food might have. Some people are concerned that we might not be adequately testing bioengineered foods for their effect on the humans who consume them.

A study by the National Research Council concluded that there is no evidence suggesting that bioengineered food is unsafe to eat (Figure 14–6). They found virtually no distinction between the risks posed by genetically engineered plants and the crossbred plants that we currently consume. Both the **Animal and Plant Health Inspection Service (APHIS)** of the U.S. Department of Agriculture (USDA) and the Environmental Protection Agency inspect bioengineered crops

Figure 14–6
A National Research Council study has found that there is virtually no difference between the risks posed by traditional crossbred foods and bioengineered foods.

produced in the United States for safety. APHIS is the leading government agency that governs the testing of bioengineered plants. It is customary that several years of field testing be done on any new plant engineered. Breeders' evaluations range from a plant's characteristics to its disease resistance. In general, any scientist or institution, whether research based or corporate, must have approval from APHIS before conducting field tests or transporting a bioengineered plant.

In addition, the USDA's Food Safety and Inspection Service inspects meat and poultry to be consumed as food, and the Department of Health and Human Service's **Food and Drug Administration (FDA)** administers the safety of all the rest of our food supply (Figure 14–7). It makes certain that any foods coming from bioengineered plants are held to the same safety standards as traditional food products. The FDA has reviewed all the bioengineered foods that have been put on the market and has not found any reason to believe that they pose a threat to the health of consumers.

Another related concern of people is that they want genetically modified foods to be labeled in stores. The labeling of biotechnology products has become a worldwide debate. In the past, consumers did not know if their food was genetically engineered. Labeling products could easily solve the fear of the unknown. Many people feel that if the product is safe, then it should be labeled, and many companies are beginning to label (Figure 14–8). These companies feel that considering consumer fears and labeling can improve their image as well as reduce fear. This would allow consumers

Figure 14–7
All meat and poultry is inspected by the Food Safety and Inspection Service.

Courtesy of USDA Image #02CS0490. Photo by Ken Hammond

Figure 14–8
Many feel that foods from GMO origins should be labeled as such to give consumers a choice.

to have the knowledge to choose between GMOs and those foods that are free of them. Those opposed to labeling contend that it would solve no real problem and would only increase fears about genetically altered food.

Environmental Concerns

Many people have expressed concerns about what effects biotechnology may have on the world around us. We share our world with many other living organisms, and many people feel that it is important to consider them when introducing bioengineered organisms into the environment. Because we coexist with these organisms, it is also believed that any environmental harm we cause will eventually affect the human population as well through outlets such as the **food chain**, water sources, and even the air that we breathe.

Bacillus thuringiensis (Bt) is a bacterium that occurs naturally and has recently become popular because of its pest control abilities. It has been used on crops such as corn, potatoes, and cotton. Public concern about the use of Bt toxins is that insect populations will become resistant to them (Figure 14–9). When there is widespread use of Bt toxins, the result is that only those insects that can tolerate the toxins will survive. When those resistant insects mate with one another and reproduce, the number of toxin-resistant insects grows, and the fear is that the Bt toxins will become ineffective as pest control. Organic growers have become

Figure 14–9
Some people are concerned that the use of Bt toxins will build up insect populations that are resistant to the toxins.

concerned because Bt sprays are one of the few insecticides that they can rely on to control the insect population that might otherwise destroy their crops.

Scientists recommend that growers use other insecticides along with Bt to help prevent the development of insect resistance. It has also been suggested that cultural control methods be used. This means taking actions such as setting up areas where Bt is not used in order to allow unaffected insects to live and breed with any resistant insects that are present. This may help decrease the likelihood that a highly resistant population will arise.

Another concern about Bt toxins is the effect that they might have on the environment or on other wildlife besides the insects that crops are being treated for. There have been claims that the Bt toxin could move through the food chain and kill beneficial insects, such as ladybeetles. There is also the concern that other organisms and wildlife that prey on insects will be harmed, such as the bird population that depends on insects for survival (Figure 14–10). In addition, other organisms that live within the soil could be negatively affected by the presence of greater amounts of Bt toxins. This could possibly lead to a buildup of plant remains in the soil if they are not properly broken down.

With the genetic manipulation of animals and particularly aquatic animals, there also comes the inevitability of "escape" from the confines of captivity because of the difficulty of confining these naturally mobile animals. If these species interbreed with their relatives in the wild, what will

Figure 14–10
It is feared that birds that prey on insects killed by Bt toxins will be harmed.

the result be? At the very least, the breeding will alter the genetic lines of the animals bred. Other risks include the danger of disrupting fragile ecosystems and causing the extinction of native species. Although the APHIS is responsible for regulating the intentional release of insects for pest management, who will guard against the accidental release of insects from those companies that have begun to farm recombinant proteins from insect larvae?

Further uneasiness has been caused by the effect that plant biotechnology may have on the surrounding environment or biodiversity. Biodiversity is known as the number and variety of organisms found within a specific geographical region. The argument surrounding biodiversities is that growing plants that are resistant to pests also takes away the food source for beneficial creatures. Opinions vary as to whether the gain of more effective food production is worth the loss of any beneficial organisms (Figure 14–11).

Additionally, some feel that biodiversities are in danger of plants that have been modified to be resistant to weeds. No one can deny the advantage of having an agricultural crop that is resistant to herbicides used to kill competitive weeds, but there is a possibility that the weeds could become resistant to the chemicals and produce a superweed. A superweed is one that is difficult to kill by current means and could possibly take over agricultural crops. This poses severe threats to many agriculture commodities, whether they are GMOs or not. The benefits of plant biotechnology range from

Figure 14–11
Concern has been expressed over the effect genetically modified organisms might have on biodiversity.

insect and disease resistance to drought and heat tolerance. Numerous developing countries where starvation is a problem could experience immeasurable personal and economic success because of the improved growth and yield of genetically modified plants. They stand to gain the most. Critics point out that the bottom line with plant gene splicing is that it is still in the infancy stages, and no one knows what effects it could have in 20 years. Many people fear the unknown and have a hard time accepting new technologies that may still have problems to be worked out.

The Department of Health and Human Service's National Institutes of Health has created guidelines for the use of genetically engineered substances in laboratories (Figure 14–12). The National Research Council has studied environmental risks and determined that there is virtually no difference in the risk level of bioengineered plants created through modern techniques and those produced by conventional methods. In addition, the APHIS makes the determination whether a plant is safe among other existing plants in the environment and whether it is as safe as plants produced by traditional methods.

The U.S. Environmental Protection Agency (EPA) puts together science advisory panels, made up of independent scientists, to address environmental concerns before and/or after a bioengineered crop is put on the market. The EPA also inspects the safety of herbicides and pesticides placed in the environment by looking at their effectiveness, their safety

Figure 14–12
The National Institutes of Health has created guidelines for the use of genetically engineered substances in laboratories.

to humans, the role that a substance will play in the environment, and the possible effects it could have on other species that it will contact.

The Altering of Animals

Critics contend that some animal species have been changed remarkably by human intervention, and they question whether the changes have been good for the species. For example, the Holstein cow of today is far different from its predecessors, even those raised as recently as 50 years ago (Figure 14–13). From breeding with select bulls only, milk production per cow has increased nearly threefold, leading to a drop in the number of cows produced and the amount of land that is required to raise and maintain them. Along with these improvements, however, have come problems: today's Holsteins have tendencies toward lameness; in addition, they are markedly less fertile than their ancestors, they are often useful as milk producers for only 2–3 years, and their genetic lines are considerably more narrow.

Also some producers are apprehensive about the increasing pressure to produce leaner, faster-growing stock of uniform size. This is of particular concern in the swine industry, where pressures such as these are leading to the abandonment of older breeds. As the stricter ideals are spread to other species, it may be that unless a particular breed of livestock is used for consumption, for its milk, or for other by-products

Figure 14–13
The Holstein of today is different than its predecessors.

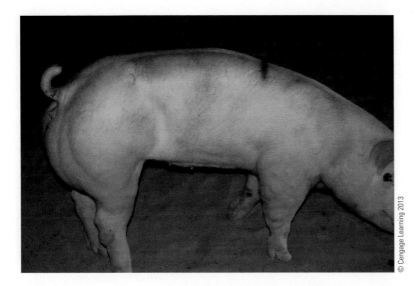

© Cengage Learning 2013

Figure 14–14
The selection of pigs for extreme muscling and leanness has led to problems.

such as wool, it will be allowed to simply die out, its "usefulness" destroyed by our quest for faster, better, and more economical products. Choosing pigs for extreme leanness has led to excitability problems during handling, and selection based on high reproductive rates or lactation rates has also led to animal welfare problems (Figure 14–14). Critics contend that once we lose the older strains of animal, they can never be recovered.

Opponents also point out that the poultry industry has seen similar effect. Today's breeds grow nearly twice as fast, produce more eggs, and can be maintained on much less feed. They contend that many traditional breeds have been lost, and today's breeds have shown a decline in genetic diversity. Some have estimated that there are fewer than 10 commercial strains of chicken currently available to North American suppliers, down from several hundred different lines at the beginning of the twentieth century (Figure 14–15). Physical abnormalities such as skeletal weakness, cardiovascular problems, and gait problems can all come from breeding for increased growth rate.

Animal welfare also becomes an issue where reproductive manipulation is concerned. Practices such as artificial insemination, embryo collection, embryo transfer, and superovulation are all used in the production of transgenic animals. In some species of livestock, these procedures require surgery or other invasive techniques that create the potential for pain. In still other species, such as poultry, the animal must be killed in order to perform procedures such as the

Figure 14–15
Some have estimated that there are fewer than 10 commercial chicken currently available to North American suppliers.

collection of eggs, sperm, or embryos. Animal rights activists say that such practices should be banned. Proponents argue that some replacement or alternative procedures have been developed for obtaining embryo or egg collection that involve nonsurgical or less invasive means, such as collection from slaughterhouses, where possible.

Genetic Transfer in Animals

Gene transfer has great potential in at least three areas: (1) for greater production efficiency of farm animals, (2) as a possible aid in the use of livestock to produce medicine, and (3) as tissue for human transplants. One concern is over the process of genetic transfer methods. **Microinjection** is a method of gene transfer in which DNA is injected into the embryo of an organism. Mice are used experimentally to determine the results of this type of gene transfer. It has been estimated that 5–10 percent of established transgenic mice lines produced by microinjection have mutations, such as severe muscle defects, sterility, shortened life span, missing organs, seizures, limb deformities, and brain damage. The majority of these animals never make it past the womb, but the approximate 25 percent that do are faced with these abnormalities. It is likely that similar problems would be found in microinjection livestock at similar rates. Proponents point out that great strides have been made and are continuing to be made in the methods used to clone and splice genes. These newer methods have far fewer detrimental effects, and it is just a matter of time before the techniques are perfected.

Scientists hope to use genetically engineered livestock to generate products such as pharmaceuticals, to provide an alternate source for organ replacement for humans in need of transplants, and to serve as models for human diseases. For example, clinical trials are already under way to test the viability of producing proteins in milk by expressing foreign genes in the mammary glands of livestock. Theoretically, these animals can produce more nutritious milk or milk that is more readily converted into cheese and other dairy products. Perhaps the greatest interest in producing transgenic livestock has come from pharmaceutical companies wanting to produce clotting factors, enzymes, and other proteins in milk (Figure 14–16).

Some consumers are concerned about the safety of milk that has been produced from genetically modified cows. They contend that the milk will not be natural, and there may be some unknown problems with its safety for human consumption.

There is a huge shortage of organs available to those human patients desperately needing transplants in order to stay alive. The pig is considered to be a likely source for

© bluewren08. Image from BigStockPhoto.com.

Figure 14–16
Research is under way to use genetic engineering to increase milk's nutrition and to produce clotting factors, enzymes, and other proteins in milk.

human transplant organs. In fact, there have already been experimental human transplant surgeries involving organs taken from pigs and other animals. In addition, genetically engineered sheep have been produced to carry a mutated collagen gene, possibly making them a model for human connective tissue diseases. Critics contend that placing foreign animal tissue into humans may have results that we now know nothing about. The concern is that using nonhuman tissues will alter a human body. Advocates counter with the argument that in the future genes from the person needing the transplant will be inserted into the donor animal. They contend that this will make the organs very much like those in the human body.

Cloning is another issue that brings about much debate. Most of the disagreement is over the ethics of cloning, a topic is covered in depth in Chapter 15.

The Governing of Biotechnology

The actions of governments around the world reflect the same philosophy of implementing rules and regulations that will provide standards and guidelines for biotechnology. A regulation simply provides a principle, rule, or law designed to control or govern conduct. Lawmakers try to keep abreast of the issues and understand the immense implications of genetic science (Figure 14–17). Eleven

Figure 14–17
Lawmakers are working to implement rules and regulations to govern genetic science.

pieces of legislation were introduced in 2001 that attempted to place a ban or moratorium on genetically modified seeds, crops, or animals for a specific length of time. State or city governments imposed all of these. These moratoriums call for the suspension of an ongoing activity for a certain amount of time. This allows the government to gain more information about the biotechnology process and have time to decide what implications it could have for citizens. In the United States, there are no formal regulations on biotechnology food products. Currently, they fall under the same standards as conventional food products. The FDA's stance is that these foods are no different from any food that must be approved for human consumption, and the same stringent requirements must be met. Nevertheless, the United States applies present food safety and environmental protection laws and regulations to biotechnology products and approves or disallows their use for consumption on the basis of the qualities of the products rather than on whether they are a derivative of genetic modification. Among the qualities measured in decisions to approve a biotechnology food product (whether plant or animal) for human consumption are its expected nutritional value, its ability to be rapidly digested, and the extent to which the biotechnology component is largely the same as other proteins present in conventional food.

GMO Labeling

The government has not enforced the act of labeling genetically modified foods to be sold in stores yet. This is a point of contention for lots of people who do not feel comfortable about buying something because they are unsure of the contents. Nonetheless, many companies are voluntarily labeling modified products, largely to gain public awareness that they are the same as traditional foods. Several large companies feel that by labeling products, they can overcome the anxiety many consumers have when buying modified foods (Figure 14–18).

As for other government initiatives, there is a **National Bioethics Advisory Commission** that gives advice to the president, although the Food and Drug Administration decides whether a particular medicine is safe. Several other

Figure 14–18
The government has not enforced the labeling of genetically modified foods.

countries are also going through the same process as the United States in trying to create safe and ethical guidelines for biotechnology, many of whom watch U.S. progress and match its decisions. Global guidelines are in the works for biotechnology food products that would be internationally enforced.

Summary

In previous chapters, we have discussed some of the reasoning behind methods such as genetic engineering and cloning. As the world population rises and the human standard of living improves, demand for high-quality agricultural products becomes greater. When this is combined with the fact that our agricultural land is disappearing, there is ever-increasing pressure to use biotechnology to improve product quality and increase productivity.

Considering the constant advances being made in the biotechnology field, these issues will definitely continue to be hotly debated as the uncertainties and potential risks are weighed against the possible benefits to society. Opposition from a portion of the public is not likely to cease until we have resolved its questions about food safety and the safety of our environment. Both sides have been guilty of using available information to their advantage. The biotechnology industry has sometimes been accused of glossing over potential risks while overstating the potential benefits of genetically modified organisms (GMOs). At the same time, opponents often exaggerate the possible risks and downplay the benefits that could come from GMOs, and sometimes they may make claims that have little or no scientific merit. Both sides must work together to make judgments based on scientific facts rather than on potential profits or on emotion, fear, or unsubstantiated claims.

Controversy brings about discussion, and it will take discussion, as well as compromise and understanding, to end up with a solution that will work for society as a whole. We all have something at stake in this situation, and if we are all able to voice our views and concerns while recognizing that there are other perspectives, there is hope that we can come to an agreement that will benefit us all.

CHAPTER REVIEW

Student Learning Activities ....

1. Do an Internet search for studies on bioengineered foods. Choose one study; report to the class on who conducted the study and what their findings were. Give your opinion on whether you think the study is valid or invalid and list the reasons why.

2. Conduct your own study. Interview students and teachers in your school and ask, What is your definition of bioengineered food? Do you feel safe eating these foods? Should scientists be genetically manipulating animals and plants if it benefits humans? Compile your research and share your findings with the class.

3. On the Internet, find a country that has banned the use of GMOs. What rationale does it offer? Do you feel that this is a valid reason?

4. Organize a class debate on the use of GMOs. Be sure that arguments for and against are based on science.

Fill in the Blanks ....

1. The primary _____ concerns within the _____ field today revolve around _____, _____ concerns, and the current debate surrounding the development of _____.

2. _____ plants has been the traditional method of improving food source _____ and _____, but it is not _____, and it involves a lot of _____ and trial-and-error methods.

3. Many genetically engineered _____ have had a _____ _____ gene introduced to _____ _____ the growth process.

4. A study by the _____ _____ _____ concluded that there is no _____ suggesting that bioengineered food is _____ to eat.

5. Many people believe that any _____ harm we cause will eventually affect the _____ population as well through outlets such as the _____ _____, _____ _____, and even the _____ that we breathe.

6. It has been estimated that _____ to _____ percent of established transgenic mice lines produced by _____ have _____, such as severe muscle defects, _____, shortened _____ _____, missing organs, seizures, limb deformities, and brain damage.

7. Scientists hope to use genetically engineered _____ to generate products such as _____, to provide an alternate source for _____ replacement for humans in need of transplants, and to serve as _____ for human diseases.

8. Among the qualities measured in decisions to approve a biotechnology food product for human _____ are its expected _____ value, its ability to be rapidly _____, and the extent to which the biotechnology component is largely the same as other _____ present in _____ food.

9. There is a _____ _____ _____ _____ that gives advice to the _____, although the _____ _____ _____ _____ decides whether a particular _____ is safe.

10. Eleven pieces of _____ were introduced in 2001 that attempted to place a _____ or _____ on genetically _____ seeds, crops, or animals for a specific length of time.

True or False

1. We have greatly altered nearly every domesticated crop plant species from its original, natural form.

2. Estimates are that 15 percent of the U.S. population suffers from food allergies.

3. A very small percentage of allergens are proteins.

4. Some consumers are concerned that bioengineered foods have not been adequately tested to determine if they are safe for human consumption.

5. The FDA has not found any reason to believe that bioengineered foods post a threat to human health.

6. Fewer companies are labeling genetically modified foods now than in the past.

7. Organic farmers are very concerned about insects becoming resistant to Bt toxins because Bt sprays are one of the few insecticides they can use.

8. Microinjection is a method of gene transfer in which DNA is injected into adult cells.

9. Biotechnology food products currently fall under the same FDA regulations as regular food products.

10. Many companies are voluntarily labeling genetically modified food products.

Discussion

1. What does the term "frankenfood" mean? What is the significance of this term?

2. Discuss why some people are concerned about food allergens being introduced through genetic engineering.

3. Why do proponents of GMOs argue that gene splicing is safer than crossbreeding?

4. Describe the roles APHIS and the EPA play in the regulation of bioengineered foods.

5. Explain how insect resistance can result from the use of Bt toxins.

6. Describe the effects Bt toxins may have on the food chain.

7. What effects have human selection had on the biodiversity of domestic animals?

8. Describe the areas of great potential for gene transfer in animals.

9. Why are some consumers concerned about the lack of labeling for GMO food products?

10. Do you believe that genetically modified food products are safe for consumer consumption? Why or why not?

CHAPTER 15

Ethical Issues in Biotechnology

KEY TERMS

ethical dilemmas
ethics
Genetic Information Non-
    discrimination Act of
    2008
pluripotent
patents
bioprospecting
biopiracy
biodiversity
sustainability

OBJECTIVES

When you have finished studying this chapter, you should be able to:

- Define ethics.
- Discuss the benefits and concerns of stem cell research.
- Explain patents on life.
- Discuss the controversy surrounding the patenting of life forms.
- Discuss the benefits and concerns of cloning.
- Describe the regulation process of biotechnology.
- Express how ethics may affect the future of biotechnology.

Ethical Dilemmas

Suppose that one day you are in the chemistry lab conducting an exercise by following the instructions given to you by your teacher. You have followed the instructions very closely, and so far you have completed all the steps and recorded all your actions in your notebook. Suddenly, you realize that the class will be over in 5 min and you still have the last step to finish. Since you know what will be the eventual outcome and time is running out, you are tempted to write up the last step as though you had completed the entire exercise. You feel that you have gained all the knowledge and experience that you can even without finishing the exercise. On the other hand, you realize that you are writing false data in your notebook and that you are not being truthful. What should you do?

One day, you discover that your friend has taken money from a fund-raising activity in order to pay for her lunch. You know that she is very poor and may not have enough money to pay for her lunch. You do not want to see her go hungry. Besides, the fund-raising has gone extremely well and has made more than enough money to pay for the trip next month. Yet you know that stealing is wrong and that the theft should be reported. Should you report the theft or just look the other way?

These two scenarios are what are known as **ethical dilemmas**. What is the correct thing to do in each instance? Logical-sounding arguments can probably be made for taking either action, and both directions can result in consequences. The study of **ethics** deals with making a decision and taking an appropriate course of action on the basis of what is right or wrong, moral or immoral (Figure 15–1). Almost every day, we are faced with making decisions about what is the right thing to do. The solution may be simple and can be made without a lot of thought, or it may be extremely complicated to arrive at the right conclusion. A concrete definition of ethics can fluctuate among individuals because of differing views on what is good, bad, moral, or immoral. Conflict often arises because of the differing values among people or even whole societies. A lot of the time, conflict comes about because people do not understand the beliefs and values of those taking

Figure 15–1
The study of ethics deals with making a decision and taking an appropriate course of action on the basis of what is right or wrong, moral or immoral.

Figure 15–2
The implications of many scientific discoveries and developments can be hard for people to understand.

the opposite view of the issue. This is why it is so essential that we make an effort to understand various views and cultural background.

Often, scientists make new discoveries or create new developments that people do not understand (Figure 15–2). Also they may solve one problem and create another problem by implementing the innovation. For example, an exotic plant may be brought in that is superb at preventing erosion but may become a very invasive weed that takes over cropland. When this happens, friction may result between the scientific community and the general public. This is why it is so very important that both sides of an issue be thoroughly discussed before reaching conclusions as to what course of action to take in releasing or implementing any new innovation. Once the facts have been debated, the ethical aspects of an issue have to be decided before any action can be taken.

Bioethics is the area of ethics that is specifically associated with biological and medicinal research. Almost everyone agrees that life (particularly human life) is most precious and anything that affects life can have serious consequences. Ethics dealing with life, especially human life, are almost as old as human civilization. One good example is the Greek Hippocratic Oath created by a physician named Hippocrates who admonished his fellow doctors to value human life and to have standards of conduct regarding the people they treated. Today, all physicians in this country take the Hippocratic Oath before they are licensed to treat people. This oath outlines the ethical treatment of patients and provides guidelines for determining the moral or ethical aspect of medical decisions.

In many areas of study, scientists struggle to weigh the potential dangers of their research against the expected rewards of their findings (Figure 15–3). Most research-and-development projects have some risks. The ethical

Figure 15–3
Scientists must consider the potential dangers of their research as well as its expected benefits.

responsibility of research scientists and those who regulate the release of research findings is to determine the risks involved and if the benefits are worth the risks. This determination is often the point of controversy.

Research in biotechnology has opened doors that some scientists believe will lead to a world without famine, disease, or genetic disorders. Despite the fact that there are numerous benefits linked to biotechnology, there are also a great number of concerns and ethical issues involved. Even though we have been genetically modifying organisms through selective breeding for thousands of years, there are some people who feel uncomfortable with genetic modification that takes place in the laboratory. They view genetic engineering as an inappropriate use of technology because of the level of control humans are exerting over the natural world. Determining the line between ethical and unethical is a complex and difficult task because ethics is governed by an individual's moral principles and guidelines as he or she perceives what is good and what is evil (Figure 15–4). People with different religious backgrounds often have conflicting views. For example, the Hindu religion holds that cattle are sacred animals and are not to be used for beef. But other religions do not share this same conviction, finding it acceptable to use cattle for food and other purposes (Figure 15–5). Those of the Muslim and Jewish faiths believe that it is wrong to eat pork and meat from animals they deem unfit.

© Radu Razvan/www.Shutterstock.com

Figure 15–4
Determining the line between ethical and unethical can be a complex and difficult task because ethics is governed by an individual's moral principles and guidelines.

Figure 15–5
The Hindu religion holds that cattle are sacred animals.

Ethics is also regulated by the facts of a particular situation. For instance, many people are against cosmetic companies using animals for product testing; however, when the situation is changed and animals are used for research to find cures for diseases, the general consensus shifts to support for animal research. The difference in opinion relates the perception of the risks and benefits of the research.

A huge task in solving an ethical problem is separating human emotions from factual information. An individual's morals, values, and religious beliefs are often very strong and are instrumental in developing the way people think. It is essential when analyzing a question of ethics to gather as much information as possible on both sides of an issue in order to make a responsible decision. Always keep in mind that ethics can vary from person to person, depending on an individual's perspective and background. The following sections discuss some of the ethical issues faced by research scientists who develop biotechnology and by the general public who are asked to accept and use the new discoveries.

Ethics in Personalized Medicine

Because tools of biotechnology are widely employed to improve human health, issues of patients' privacy and medical ethics are important to consider. For example, patients should give informed consent when submitting biological

Figure 15–6
Thought must be given to deciding which individuals or groups should have access to genetics information.

samples that may yield genetic information. Thought must be given to deciding which individuals or groups should have access to genetic information (Figure 15–6). Only patients? Family members? Doctors? Insurance companies? Employers? In an interesting case in Iceland, the government's socialized medical records were sold to a research organization, deCODE genetics, that combined patient medical histories with family history data. Citizens were then asked to give genetic samples as part of routine medical examinations, with the ability to "opt out." However, because Iceland was settled by a small founder population, most of today's citizens share common ancestors. This makes it an ideal human population for studying the effects of specific genes on health. However, even if a citizen opted out of the deCODE study, it would be easy to infer the person's genetic background by looking at his or her relatives. The deCODE project raised ethical questions about informed patient consent, patient privacy, data sharing between governments and private companies, and ownership of genetic information.

In the United States, the **Genetic Information Nondiscrimination Act of 2008** was passed to prevent discrimination against individuals in regard to health insurance or employment, based on their genetic information. In addition to medical providers, a number of personalized genomics companies have started in the United States in recent years. These companies aim to provide individuals with genetic screening for single nucleotide variations that may be correlated with medical traits. Ethical issues around personalized genomics include data security (will databases with genomic data ever be hacked?) and whether customers can reasonably interpret the statistics that correlate a specific genetic trait with the likelihood of developing a particular medical condition. There are currently paternity testing kits that may be purchased at drugstores, and some genomics companies aim to provide similar home test kits for genetic diseases. Proponents say that everyone has a right to learn about their own genetic variations, but opponents claim that the average person may be unable to appropriately interpret genetic test results and should be able to take genetic tests only under medical supervision.

Ethics in Stem Cell Research

One area of biotechnology that attracts a lot of attention is stem cell research. Stem cells are cells that grow and differentiate into the various tissues that make up the assorted organs and systems of an animal's body. Because they are able to develop into many different types of cells, stem cells are **pluripotent**. Scientists are now able to culture stem cells in the laboratory. If they can unlock the process of cell differentiation, they can grow custom-made tissues (Figure 15–7). For example, adult stem cells might be taken from a person who has been badly burned, and from these stem cells new skin could be grown. In addition, transplant organs could be

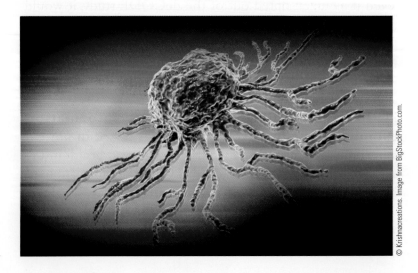

Figure 15–7
If scientists can unlock the powers of cell differentiation, they can grow custom-made tissues.

grown from stem cells that came from the person needing the transplant. Since the transplant cells and tissue are the same as those cells of the person's body, rejection of the new organ would be unlikely.

Although the results remain to be seen, scientists look to stem cell research as a cure for degenerative, chronic, and acute diseases, such as Parkinson's or diabetes. Millions of people bear the burden of disease, and almost every individual is directly affected by disease in their lifetime, either personally or through the suffering or death of a loved one. By better understanding how cells differentiate, we may be able to determine what goes wrong in cells that develop improperly and cause genetic diseases.

Stem cell research may provide the weapons to eradicate debilitations such as those resulting from spinal cord injuries (Figure 15–8). Injuries to the nervous system are especially difficult to treat because they ultimately cause the impairment or loss of nerve cells. Adult nerve cells cannot divide to replenish those that are damaged or killed. Diseases such as Parkinson's, Alzheimer's, and some types of sclerosis all attack the nervous system. Spinal cord injuries, strokes, and aneurisms all affect the nervous system, harming or destroying nerve cells that cannot be replaced. Stem cell research provides the prospect of generating fresh nerve tissue that can divide and restore damaged nerve cells.

Much of the controversy surrounding stem cell research is over the process of gaining stem cells from human embryos (Figure 15–9). Stem cells may be derived from immature human embryos or aborted fetuses. Once they are cultured, these cells can replicate indefinitely. Opponents of stem cell research claim that it is immoral to use human embryos or fetuses for research regardless of the potential benefits. They point to obtaining adult stem cells from willing participants as an alternative means of procuring cells that may differentiate into many types of tissue. However, it is possible to create human embryonic stem (hES) cell lines without destroying embryos. In a technique that is widely used to screen for genetic diseases, a single cell may be taken from an eight-cell embryo without damaging the embryo. This single cell may be cultured into an experimental cell line.

Scientists assert that embryonic stem cells offer more versatility and are more readily available than adult stem cells. One of the greatest ethical questions in biotechnology is

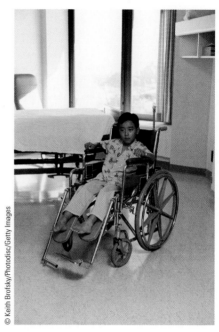

© Keith Brofsky/Photodisc/Getty Images

Figure 15–8
Stem cell researchers may discover treatments for debilitating conditions, such as spinal cord injuries.

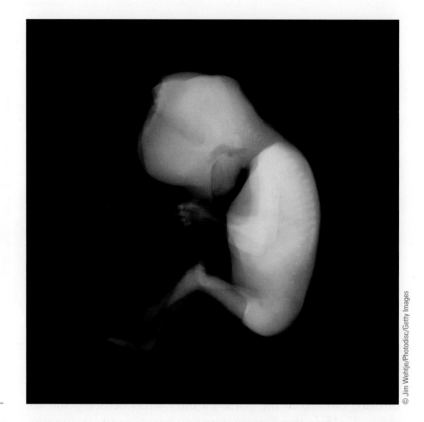

Figure 15–9
Stem cell research is controversial because stem cells can be derived from human embryos.

how life is defined or when life begins. There are many different opinions and perceptions as to the best definition of life. Many consider life to begin at conception, others say that life begins when the fetus is formed, still others believe that life begins when a fetus can live on its own, while yet others contend that life begins at birth. This type of ethical question is difficult to answer through scientific research. However, the perception of when life begins is a central point in determining whether hES cells should be used for research.

A highly publicized debate has existed during the past decade over the use of federal funding of stem cell research. A previous decision to ban federal funding of new projects was overturned in 2009, and new research projects are currently being funded using taxpayer dollars. Advocates of this research claim the government funding is necessary to further our understanding of stem cells and how they can be used to promote human health. Some opponents want all research using stem cells from human embryos halted because they believe that human life is so precious and that the life of an embryo is just as valuable as the life of a person.

Ethics in Cloning

Many consider cloning to be the most controversial area of biotechnology, with human cloning being the more specific area of concern. Human cloning has been the subject of science fiction for decades, but it has been only in recent years that this technology has become a real likelihood. As discussed in Chapter 8, cloning came to the forefront of the news in 1996 with the presentation of the cloned sheep Dolly. This milestone in scientific research was met with tremendous curiosity as well as a backlash of concern. The apprehensions ranged from religious concerns to questions of animal welfare. However, scientists involved in this field of research argue that the potential for genetic improvement in livestock and other species outweighs the associated risks.

The potential benefits for cloning livestock are numerous and fascinating. A producer with a prize cow that can no longer reproduce can take advantage of cloning to ensure that her genes influence the herd for years to come. Since the successful cloning of mules and horses, some people have even dreamed of using cloning in the horse industry. Imagine being able to clone a gelding that has exhibited tremendous talents in reining, cutting, or any other area of performance (Figure 15–10). Cloning him could produce a stallion that might pass on those desirable traits to offspring. What if producers could guarantee that each animal they sent to slaughter would grade choice or prime? The possibilities of cloning are immense, but at the same time, little is known about this developing area of science, and there may be valid concerns. For example, critics point out that although we have had success with cloning many animals, we still know relatively little about the long-term characteristics of cloned animals. They point out that we cannot be certain how cloned animals grow, develop, and age until we have a population of many hundreds of cloned animals to study. Critics question the ethics of producing this many animals through a process that we know so little about. Proponents argue that this is the case with any animal bred since a purpose is an "unknown" until several generations have passed. They further contend that the animals that have been cloned appear to be, for the most part, as normal as any other animal.

Perhaps the greatest objection to cloning is that the technology will lead to human cloning. Several different surveys indicate that the majority of people believe that

Figure 15–10
Through cloning, it may be possible to reproduce a valuable gelding.

Figure 15–11
Many people object to cloning technology because they fear that it will lead to human cloning.

to pursue the cloning of a human is immoral and unethical (Figure 15–11). Some argue against human cloning because a human clone would exist as a result of a lab experiment. Others are concerned about the incomplete knowledge of the cloning process and point to the many failures and problems encountered in animal cloning. Still others believe that creating clones is "playing God" and fear religious repercussions.

Proponents of human cloning believe that cloning would offer ways to improve the human gene pool and eliminate abnormalities and maybe even create super-humans with perfect health and extreme intelligence. They envision a population as athletic as Michael Jordan and as smart as Albert Einstein. In addition, they point out that many genetic disorders can be completely wiped out through the process of cloning. This is closely related to the debate over stem cell research because the only real application of stem cells for improving human health involves cloning of a human being in order to provide stem cells that can be used to treat a disease in that individual.

Opponents of this argument point out that the process could just as well produce some of the world's most infamous criminals. They also contend that famous people were also a product of their environment and that the way they grew up contributed as much as genetics. Perhaps the most compelling factor in the debate is the image of the human genetic experiments conducted by the Nazis prior to and during World War II. This image is a reminder of the horrors associated with that time.

The cloning of animals requires the use of many embryos before the process is successful. Some people point out that it would be wrong to destroy human embryos in order to obtain a clone. This argument also centers around the debate over when life actually begins.

Who Should Benefit from Biotechnology?

With all of the tremendous advantages of biotechnology, it stands to reason that there are huge amounts of money to be made through the development and implementation of the technology. Currently, there is much debate in relation to the profits of large corporations responsible for advances in biotechnology (Figure 15–12). Many questions surround the ethics of whether a select group of companies should get rich off of information and technology that could spare millions from hunger and suffering. However, biotech business is high risk, and there are long time lines involved in bringing a product to market. To pay for the research and development of technologies, which often costs hundreds of millions of dollars, companies must be able to pay their employees and make some profit. But who should decide how much profit is justified and fair to all?

The ethical predicament in this case is that many of the countries in the greatest need cannot afford the technology needed to get them out of the vicious cycle of poverty and starvation. Through the great advances in plant and seed technology brought about by biotechnology, many drought- and disease-stricken countries could see great improvements in health and quality of life (Figure 15–13). Governments and charitable organizations play a role in making technologies, such as genetically modified seeds or biomanufactured pharmaceuticals, available in developing countries. Companies have also come up tiered pricing strategies that charge more to customers in the developed world and less to customers in developing countries. However, if government regulations were imposed on large corporations to share with other countries the wealth and expertise gained, would these companies still devote as much money and effort to research

Figure 15–12

The profits of large corporations responsible for advances in biotechnology can be controversial.

Figure 15–13
Biotechnology could be a real boost to poverty-ridden countries if they could afford the technology.

and development? Is it ethical to stifle the free enterprise system by imposing such restrictions? Only time will tell if and how the government's actions will shape the future effects that major corporations have on the rest of the world.

Other ethical questions are the following: Is it ethical to give the technology (or products of the technology) to some countries and not to others? If this is ethical, how do we decide which countries get the technology and which do not? If countries are given the technology (or are sold the technology), could these countries then sell the technology? If they are not allowed to sell the technology, how will this regulation be enforced? The answers are not easy, and no matter what is decided, the controversy will probably continue.

Patents on Life

Figure 15–14
A patent gives an inventor exclusive rights to use or profit from the invention.

One of the many great ethical concerns of biotechnology is the emerging dispute about patents on life. **Patents** are government grants that provide an inventor with the exclusive right to use, sell, and manufacture an invention for a set period of time (Figure 15–14). However, within the past few decades, as a result of the advances in biotechnology, this grant has also extended to life forms and genes. One of the requirements of obtaining a patent is to prove that the product has never been made before, involves a nonobvious inventive step, and serves some functional purpose. Until the 1980s, these conditions excluded living organisms. Formerly, living organisms were regarded as discoveries in nature, which were therefore unable to receive a patent. However, in 1980, this all changed in a monumental United States Supreme Court case, *Diamond v. Chakrabarty*, where the Court ruled that a living organism, in this case a bacterium that could digest oil, was patentable. The Court's rationale was that the genetically modified bacterium was not found in nature but had been engineered by scientists (Figure 15–15).

Some people still question whether naturally occurring living organisms or genes can be properly regarded as human inventions, either before or after genetic modification. This is a tough ethical question. The answer waits in the principle of whether people believe that the manipulation of genes is the creation of life. Patents on life ensure that a research group owns the sole rights associated with the "created" organism or genetic product. Recently, synthetic biologists, led by J. Craig Venter, have demonstrated that it is possible to make synthetic chromosomes in the lab and genetically reprogram cells, switching their species identity. Have they created new life even though they used an existing cell as the foundation for their efforts? Should anyone own a patent on a form of life? What if a cancer patient has surgery and the tumor tissue removed by the surgeon later becomes a valuable patented cell line, leading to the development of an anti-cancer drug worth millions of dollars. Should the surgeon who harvested and cultured the tumor tissue be allowed to patent the patient's cells? Under what circumstances would this be ethically okay? Should the patient receive some monetary benefit? To whom does the valuable cell line belong?

Patenting allows select research groups to take control of and exploit organisms and genetic material as private property that can then be licensed or sold to producers, breeders, scientists, or doctors at a cost set by the research group. Research groups argue that patenting is fair if they spend the money and resources to develop the technology. They contend that the only way to fund expensive research and to develop new technology is through the royalties they earn under the patenting process.

For many years, companies have been able to patent genes found in nature as well as those genetically engineered in the lab. For example, mutations in two genes have been linked to susceptibility to breast and ovarian cancer: BRCA-1 and BRCA-2. In 1994, Myriad Genetics filed

© D. Falconer/Photolink/Photodisc/Getty Images

Figure 15–15
In 1980, the Supreme Court ruled that a living organism (a bacterium that could digest oil) is patentable.

for a patent on the BRCA-1 gene sequence that would be the basis of a genetic test that the company wished to develop. Recently, the patent was challenged by a group led by the American Civil Liberties Union (ACLU) and Public Patent Foundation based on the idea that this gene sequence occurs in nature and belongs to all of humanity. In 2010, the United States Department of Justice weighed in on the issue and overturned long-running practices by the United States Patent and Trademark Office. Basically, the government has stated that isolated and unmodified human genes are naturally occurring and are therefore unable to be patented.

The Myriad Genetics case is ongoing, but the impact has already been far reaching. One of the key bioethical issues of our time is transparency and data sharing. Some feel that we are ethically obliged to share technologies that have a great positive impact on all of humanity, as all people have a basic human right to enjoy the benefits of scientific progress. Patents can be a good tool for sharing technological advances, as inventors must disclose all background information and explain exactly how their invention works. Patents also allow researchers to license and share technologies responsibly. To maximize the social benefits of new technologies, scientists need to work with political, legal, and economic experts to develop fair methods of distributing information.

Should this patenting extend to human life as well? Some contend that if patents on human life were ever approved, it would mean that human body parts or even human clones could be patented. The moral decision of humans being patented must consider whether human life is indeed sacred or whether it is a commodity to be sold and traded. Much controversy still surrounds this topic, as its outcomes have yet to be determined.

Biopiracy

The patents-on-life issue leads to another ethical predicament concerning biotechnology. For thousands of years, people have been collecting plant species and other living organisms while on voyages or military expeditions. A similar term for today's generation is **bioprospecting**, which is the process of collecting biological samples that can help

Figure 15–16
Bioprospecting is the process of collecting biological samples that can be helpful in scientific research.

medicinal and scientific research (Figure 15–16). However, some consider this to be what is known as **biopiracy**, which is a relatively new term in the field of genetics.

Biopiracy refers to the illegal commercial development of naturally occurring biological material, such as plant substances or genetic cell lines. Biopiracy sometimes occurs by a technologically advanced country or organization without fair compensation to the peoples or nations in whose territory the materials were originally discovered. Tropical countries contain some of the richest genetic diversity on the planet, and many industries send scouts to these countries to seek out valuable organisms or plants. Once the scouts find these valuable species, they take samples back to their laboratories, where scientists isolate active ingredients or genetic sequences and patent them as their own inventions. Some companies are trying to patent and market remedies or benefits from plants that they are calling new inventions, when indigenous people have known about them for centuries. Biopiracy is considered by many to be an immoral use of traditional and biological heritage. The ethical concern is that the precious **biodiversity** in these countries could be altered because of the overuse and illegal seizing of unique species (Figure 15–17).

Furthermore, many species are disappearing faster than they can be protected or even studied. Several organizations and individual governments are encouraging sustainability of these unique species, which basically means keeping them in existence for future generations. **Sustainability** is a moral

Figure 15–17
The ethical concern of many is that biopiracy could reduce biodiversity in tropical countries.

obligation of people all over the world to preserve as much of our natural earth as we can so that future generations will be able to enjoy it as well.

Summary

Ethics is a very delicate subject; there is no clear set of morals that all people follow. Even so, ethics is the basis for all civilized societies, and almost everyone in the world has some set of ethics that he or she follows every day. Since ethics varies greatly from one person to the other, this is where the question of what is and is not ethical becomes difficult. People have been using ethics to guide themselves since time began, and now more than ever in the face of such advanced technology, people must draw on their ethics. Humans have a great responsibility to this and future generations to incorporate ethics into the decisions made in biotechnology, using life science to promote human health, sustainable agriculture, and environmental biodiversity. From genetic testing to creating genetically modified organisms to patenting elements of life, there are many opportunities to employ ethical debates on the influence of science on society. By understanding the advantages and disadvantages of biotechnology, people can become better informed and have a broader sense of their ethical accountability. In order to preserve the quality of life we all enjoy, people must become educated about the advantages and disadvantages of biotechnology and make their own moral decisions. Inventing sustainable biotechnology will require the incorporation of ethics into every aspect of life. Each of us is morally and ethically bound to our society and environment to try to make it the best it can be. Biotechnology has immeasurable possibilities to create a world free from starvation, disease, and many other atrocities now plaguing our world; however, in order for this to successfully happen, decisions must be grounded in ethical values. The ethical values you hold can shape the world.

CHAPTER REVIEW

Student Learning Activities

1. Search the library or Internet for reports on ethics in biotechnology. Research the issues found in the report and make notes of the points that are important to you. Share these points with the rest of the class.

2. Using your imagination to suppose that you are the president of a large corporation responsible for wonderful advances in biotechnology, create a mission statement for your company that states your ethical views on procedures in biotechnology.

3. Compare and contrast ethical and nonethical advances made in biotechnology.

4. Based on your own ethical beliefs and opinions about the controversies surrounding plant and animal cloning, write a report defending your position.

5. Use the library or Internet to research a plant or species from another country that has been illegally used in biotechnology and determine two methods for preventing this illegal use.

6. Work together in a group of five (one judge and two sets of lawyers) and choose a topic in biotechnology where ethics is an issue. One set of lawyers will be supporting the issue, while the other set will be opposing it. Use the text and other research to make your case to the judge and then swap sides.

Fill in the Blanks

1. The study of _____ deals with making a _____ and taking an _____ course of action based on what is right or wrong, _____ or _____.

2. _____ is the area of ethics that is specifically associated with _____ and _____ research.

3. The _____ Oath outlines the ethical treatment of _____ and provides _____ for determining the moral or ethical aspect of _____ decisions.

4. In the United States, the _____ _____ _____ _____ of 2008 was passed to prevent _____ against individuals in regard to _____ _____ or _____ based on their genetic information.

5. The _____ of when life _____ is a central point in determining whether human _____ stem cells should be used for _____.

6. _____ of human cloning believe cloning would offer ways to improve the human gene pool and eliminate _____ and maybe even create superhumans with perfect _____ and extreme _____.

7. _____ are government grants that provide an _____ with the exclusive right to _____, _____, and _____ an invention for a set period of time.

8. In _____, the United States Supreme Court ruled on the case, *Diamond v. Chakrabarty*, where the Court ruled that a _____ _____, in this case a bacterium that could digest oil, was _____.

9. _____, which some people call _____, is the process of collecting biological samples that can help _____ and _____ research.

10. _____ is a moral obligation of people all over the world to _____ as much of our natural earth as we can so that _____ _____ will be able to enjoy it, as well.

True or False ....

1. All societies on earth share the same ethical values and beliefs.

2. Most research-and-development projects have some risks.

3. We have been genetically modifying organisms through selective breeding for thousands of years.

4. One of the biggest tasks involved in solving ethical problems is separating human emotions from factual information.

5. Personalized genetic screening can be performed only in hospitals.

6. Pluripotent cells are cells that can develop into any tissue type.

7. Federal money cannot currently be used to fund embryonic stem cell research.

8. Many people consider cloning to be the least controversial of all biotechnological research.

9. Current cloning technology requires the use of many embryos to produce one clone.

10. The United States government has decided that unmodified human genes can be patented by biotech companies.

Discussion ....

1. In what circumstances do ethical dilemmas develop?

2. Explain why it is sometimes difficult to determine what is ethical and what is unethical.

3. Why is Iceland an ideal location for studying the effects of genes on human health?

4. Why are some people opposed to the development of home genetic tests that would be available in drugstores?

5. Describe the ethical debate surrounding the use of embryonic stem cells.

6. What are two arguments for and two arguments against cloning animals?

7. Are you in favor of human cloning or against human cloning? Why?

8. What justification do biotech companies give for charging high prices for their products and services?

9. Do you think that it is ethical to patent living things? Support your argument.

10. Explain how biopiracy may affect developing nations.

CHAPTER 16

Careers in Biotechnology

KEY TERMS

community college
graduate school
on-the-job training
doctor of philosophy
    degree (PhD)
bachelor's degree
master's degree
associate's degree
technician
good manufacturing
    practices
Food and Drug
    Administration (FDA)
marketing
Supervised Agricultural
    Experience Program
    (SAEP)

OBJECTIVES

When you have finished studying this chapter, you should be able to:

- Analyze the career opportunities in biotechnology.
- List the broad categories of job classifications that deal with biotechnology.
- Discuss the amount of education needed for different jobs in biotechnology.
- Explain some of the characteristics associated with various jobs in biotechnology.

Making a Career Decision

Now that you are in high school, you may have already considered what you want to do with your life. The choice of what career you enter is one of the most important decisions of your life. Remember that a large portion of your adult life will be spent in a job setting. If you are not happy with your work, it will be difficult to have a fulfilling life. There are many considerations in determining your career path. The financial rewards of any job are usually directly tied to the amount of preparation it takes to secure and succeed in that job. This may mean **community college**, a 4-year college, **graduate school**, or **on-the-job training**. The choice is up to you, and you must decide among a very complex assortment of job opportunities and qualifications.

Of all the activities you have engaged in, what do you enjoy doing the most? How much time and effort are you willing to spend in obtaining training or an education to get the job you want? Are you willing to leave your home area for employment? How important is making a large salary? Do you enjoy working in an office setting or outdoors? The best way to answer these questions is to get as many experiences as possible in career settings and to ask working professionals questions about their jobs. Look for internships, volunteer positions, and job-shadow opportunities in fields that interest you. It may be helpful to consult a guidance counselor or teacher in searching out these experiences.

In any career path, the ability to problem solve, follow written instructions, communicate clearly, and work well with other people is critical. In biotechnology careers, these considerations also hold true. As we have explored the many aspects of biotechnology research currently making an impact on health care, agriculture, and the environment, perhaps you have considered working in one of these areas. Does the development of an insect-resistant plant intrigue you? How about the ability to control human diseases with fruits and vegetables? If the intricacies of biotechnology interest you, it may be an option you wish to consider for a future career (Figure 16–1). As research efforts in the field of biotechnology increase, the need for well-trained individuals will increase as well. The career possibilities are as broad as the field of biotechnology.

© FikMik/www.Shutterstock.com

Figure 16–1
If you enjoy science, you may wish to consider a career in biotechnology.

If you have studied the chapters in this text, you know that biotechnology is part of most areas of biological science. For example, major advancements in the biotechnology industry have been made in such areas as drug development, human and animal nutrition, agricultural chemicals, and environmental protection. The United States currently leads the world in the research, development, and commercial use of biotechnology products. Dramatic increases have been seen in the creation of biotechnology companies and jobs over the past decade. Careers in this field are not just limited to scientist and researchers. Jobs are available for individuals with all levels of training, from a high school diploma to a **doctor of philosophy (PhD) degree**. In this chapter, you will discover some of the jobs that are available in the field of biotechnology and the educational background and experience needed for those jobs.

The number of jobs available in the biotechnology field is steadily growing. To prepare for a career in biotechnology, you should learn as much about biology, chemistry, and life sciences as possible. The only limits in this industry are your level of education, interests, and determination to succeed. Jobs can be found at all educational and skill levels, from high school graduates to those with master's or PhD degrees. Careers are also available in many different disciplines of biotechnology. Whether you are interested in plants, animals, or even human research, there are companies and jobs that will fit your interests.

In the next sections of this chapter, several example biotechnology careers are discussed. While these are by no means all the jobs in biotechnology, they represent some of the types of jobs that you might think about as a career. One great thing about choosing a biotechnology career path is the ladder-like quality of career opportunities. There are many steps on the biotechnology career ladder, with places to jump off and gain work experience, then return in later years to continue education. Many employees in the biotechnology industry start as lab assistants with an associate's degree and work toward their bachelor's degree while employed. Employers will often invest in a good employee's education, even sending bachelor's degree employees back to school for PhDs. So, if a career in biotechnology sounds appealing, there are many lifelong strategies for success.

Research and Development

Biotechnology researchers are at the forefront of new scientific discoveries. All the advancements made in the field of biotechnology have come about as a result of people who find answers to problems or questions through systematic scientific research (Figure 16–2). Many of these researchers focus their work on plants, animals, viruses, and bacteria to discover the genetic codes that dictate the biological processes of different life forms. These researchers may be involved in finding out how genetically modified organisms perform under

Figure 16–2

All the advancements made in modern biotechnology have come about as a result of scientific research.

different conditions. As you can imagine, there are almost limitless areas of biotechnology research that need to be done in the future.

Research scientists work in teams with members having varying degrees of expertise and training. As your level of education and technical ability increases, the more independent and creative work you will be allowed to perform as a member of a scientific research team. In some regions, especially nearby biotechnology centers of industry, there are specialized technical training programs that lead to 2-year associate's degrees or certificates. Two-year programs are offered through community colleges and are usually designed to serve local industry workforce needs. Graduates of associate's degree programs are sought for lab assistant positions that emphasize repetitive or specific tasks under the supervision of a senior researcher. Research technicians work more independently than lab assistants and commonly hold 4-year college degrees in a science, technology, engineering, or mathematics (STEM) discipline.

To become a research scientist, you will need several years of education beyond a basic 4-year college degree (Figure 16–3). Most scientists earn a PhD in a specialized area, such as genetics, nutrition, cell biology, microbiology, or

© Manzrussali/www.Shutterstock.com

Figure 16–3
To become a research scientist, you will need several years of education beyond a basic 4-year college degree.

biochemistry. This can take up to 4 years of graduate school after completing a **bachelor's degree** and usually involves getting a **master's degree** before entering a PhD program.

As discussed above, scientific research and development requires many well-trained people working together. Every member of a research team is vital to the achievement of research goals, and lab assistants are valued, well-paid members of the team. Some entry-level jobs in biotechnology that you may consider as a first step on this career ladder include the following.

Glass Cleaner

While this title may seem trivial, it is an extremely important job that requires basic knowledge and skill (Figure 16–4). Most laboratory experimentation involves the use of test tubes, Petri dishes, and an assortment of different types of glassware. Once the glassware has been used, it must be thoroughly cleaned and sterilized. If the proper sanitation is not exercised, unsterile glassware could contaminate the next experiment. This job requires a high school diploma along with some on-the-job training.

Greenhouse Assistant

If you like working with plants, you might like a career as a greenhouse assistant. In this job, you would be responsible for growing and caring for plants in a greenhouse

Figure 16–4

A glass cleaner makes sure that all the glass instruments and equipment are kept clean and sterile.

(Figure 16–5). The plants may be used in a scientific study, and this would require extreme care to grow and maintain the plants according to strict guidelines developed by the research scientist. This job requires a high school diploma with some on-the-job training and experience.

Plant Breeder

A plant breeder is responsible for developing and implementing plans and programs for mating various lines and strains of plant genetics. Plant breeding usually involves working with a research scientist who has a goal of producing a particular new plant line that may be developed into a new variety (Figure 16–6). The scientist and plant breeder may be working with plants that have been genetically modified. If you like to work on challenging projects that require you to work both indoors and outdoors, a career as a plant breeder might be for you. Educational requirements are at least a bachelor's degree in genetics or a closely related field.

Laboratory Assistant

Working under the direction of a research scientist, a lab assistant sets up equipment for experiments, collects data, assists in analyzing data, and may even be involved in interpreting and reporting the results of a study (Figure 16–7). By working as a lab assistant, you would be involved in many aspects of research, and you could be on the cutting edge

© Photodisc/Getty Images

Figure 16–5
A greenhouse assistant takes care of plants that will be used in scientific experiments.

© Steve Cole/Photodisc/Getty Images

Figure 16–6
A plant breeder works with various plant lines to develop new varieties.

of biotechnology. This career requires at least an **associate's degree** in science and some on-the-job training, though the majority of lab assistants have a bachelor's degree.

Media Prep Technician

This job requires preparing experiments for research scientists. Although somewhat similar to the job of a lab assistant, a media prep technician's work is more specialized (Figure 16–8). Your work would include preparation of solutions and bacterial and cell culture media, such as agar. It would also include operating balances, pH meters, stir plates, pumps, filtration equipment, open mix tanks, and autoclaves. Not only must the media be proper for the type of organisms to be grown in it, but the media and growth must be monitored and contained in order to prevent contamination of the environment. This job usually requires a bachelor's degree, although with enough experience and on-the-job training, you may be able to succeed with an associate's degree.

Animal Handler

Biotechnology in animal science requires the breeding and raising of laboratory animals for a variety of reasons, including the characterization of gene "knockouts" and other genetic mutants. Someone who understands the needs of the animals must care for them. Not only must the animals be fed, watered, and sheltered, but they must also be in proper condition for the purpose intended (Figure 16–9). For example, if the animals are to be used in a breeding program where eggs or embryos are extracted from the females, the animals must be kept in proper condition. This means that they must be fed the proper ration to keep them in the proper condition, monitored for their estrous cycle, and so on. This job requires at least a high school diploma and may necessitate two or more years of college.

Animal Technician

An animal **technician** may have the responsibility of collecting semen or flushing embryos from donor cows. He or she may take blood samples or palpate cows to determine if they are pregnant. Laboratory animals may have to be given injections, or blood samples may need to be collected (Figure 16–10).

ARS/USDA #K10023. Photo by Peggy Greb

Figure 16–7
A lab assistant helps prepare the lab and may help in the collection and analysis of data.

© Photolink/Photodisc/Getty Images

Figure 16–8
A media prep technician prepares media for growing cells, microorganisms, or other living things.

Figure 16–9
An animal handler sees that laboratory animals are well cared for.

Figure 16–10
An animal technician may give injections or take blood samples.

In general, this job involves technical skills that go beyond those required of an animal caretaker. At the minimum, an animal technician must have a high school diploma and formalized on-the-job training. Often, this position may require a bachelor's degree in animal science or a similar degree.

Technical Writer

Once a research study is completed, someone must write a report detailing why the study was done, how the study was conducted, and the conclusions and recommendations made from the findings. The research scientist doing the study usually writes the initial report and presents the findings to other scientists. If the study is to be distributed beyond the scientists, a technical writer may write the report in a manner that can be understood by the public. This job requires a background in writing composition and English skills. The typical educational level required is a bachelor's degree.

Biostatistician

In almost all scientific research, the data must be analyzed using statistical tools. These tools or tests tell a scientist whether the treatment he or she has administered has made a difference. For example, a scientist may have developed a new variety of corn that is genetically modified to resist insects. He or she may plant a field with the new variety and plant another field using a similar variety that has not been genetically modified. Once the corn has matured, the scientist will collect data. In this case, the data would be the amount of insect damage sustained by the corn in both fields. If the genetically modified corn has fewer kernels of corn with insect damage, the scientist might conclude that the new variety is effective in resisting insects. However, there is always the possibility that the difference may have occurred by chance. A biostatistician can compute statistical tests that can tell the scientist the probability of the difference happening by chance. If you are good at math and enjoy solving complex problems, a career as a biostatistician may be a good path for you. Usually, this job requires at least a master's degree in statistics.

Manufacturing and Production

Once the research has been conducted and approval has been granted to use a new biotechnology, it is up to the manufacturing industry to make a product commercially available. If research and laboratory work does not interest you, this particular area of biotechnology may be more appealing. Additionally, the background education for jobs in manufacturing and production is not necessarily limited to education in science. These jobs are open to a wide range of experiences, from business administration to electronics. Some possible jobs in this area are discussed below.

Product Development and/or Manufacturing Engineer

An engineer usually plans how the product will be manufactured or produced. This includes the design, construction, and operation of machinery or devices that can be used

Figure 16–11
Engineers develop plans for the manufacturing of a product or technology.

to take raw materials and turn them into saleable products (Figure 16–11). For example, suppose that the new variety of genetically altered corn that was discussed in the previous section is approved for commercial use. The corn seed will have to be processed and packaged in order to send out to producers. The seed has to be cleaned of debris, weed seed, and other contaminates. It then has to be put into bags in precisely the proper amount. If the bag is supposed to hold half a bushel, a machine must fill the bag with exactly half a bushel and stop. The bag has to be closed, labeled, and placed on pallets. All this takes machinery that has to be designed, built, and maintained by an engineer. An engineer has at least a bachelor's degree. If you like math and enjoy designing things, this may be an interesting career for you.

Manufacturing Technician

This job involves the running of assembly lines and other aspects of manufacturing. Machinery has to be maintained, work flow has to be monitored, and problems have to be solved. Biomanufacturing of pharmaceuticals is a tightly regulated process that must follow **good manufacturing practice** and all **Food and Drug Administration (FDA)** guidelines. A manufacturing technician keeps the process moving smoothly. This job usually requires an associate's degree.

Instrument Calibration Technician

All along the process, from receiving raw materials to shipping out finished products, instruments have to be calibrated (Figure 16–12). This means making sure that scales weigh accurately, thermometers accurately measure temperatures, electronic sensors work properly, and various types of electronic and mechanical measuring devices operate as they are intended. The minimum amount of preparation for this job is an associate's degree and a lot of on-the-job training.

Marketing and Sales

A good **marketing** and sales department is the lifeline of any company. If you are interested in the business side of biotechnology, marketing may be just the job you are

Figure 16–12
An instrument calibration technician is responsible for making sure that instruments do what they are supposed to or measure accurately what they are intended to measure.

looking for. Companies are always in need of skilled employees to analyze new products, potential customers, and market trends. These employees are responsible for the sale of a company's products or services and are also essential to the preparation of research proposals, customer relations, and technical support. Job opportunities include the following.

Market Research Analyst

Anytime a new technology or product is developed, there is a question as to whether there will be enough interest in the new item to justify manufacture and sales. A market research analyst determines how much of the product or technology can be sold and at what price (Figure 16–13). On the basis of the results of the research, a decision can be made on whether to proceed with marketing. The requirement for this job is a bachelor's degree.

Marketing Specialist

This person creates a plan to get information out to potential buyers of the product or process. The marketing specialist determines what approach is best to get the attention of those who would use the technology or product (Figure 16–14). This may include ads in magazines, radio advertisements, flyers, or other means of explaining the advantages and merits of the new development.

Figure 16–13
A market research analyst determines how much of the product or technology can be sold and at what price.

Figure 16–14
A marketing specialist creates a plan to get information about a product to potential customers.

Sales Representative

Once the product or technology is available for sale, a sales representative is responsible for getting retail distributors interested in selling the product and for getting the product to the distributors. A sales representative must thoroughly know all aspects and characteristics of the technology or product and be able to answer any questions. If you like meeting and working with people, this might be a job for you. This job requires a bachelor's degree along with some on-the-job training.

Customer Service Representative

Occasionally, a problem may arise with the new product or technology. A sales representative is the go-between that helps solve the problem for the company. Company leaders know that to stay in business, customers must be pleased with their product. The customer service representative may also teach the proper use of the product as well as help in the maintenance of the item. A bachelor of science degree in a technical or scientific field is required for this job.

Careers in Natural Resources and the Environment

If you enjoy the beauty of nature and like to be outdoors, you might consider working in natural resources, an area that will become increasingly more important as time goes on. As mentioned several times in this book, the population keeps getting larger, and the environment stays the same size. This means that we will have to do a better job of protecting our natural resources and the environment than we have done in the past. To do this will require many specially trained and educated people in different areas of biotechnology (Figure 16–15). All aspects of agriculture, particularly those that use new biotechnology, must be monitored to ensure that the environment is properly protected.

Most of the careers available in the area of natural resources and the environment will require a 4-year degree. You would need a sound background in botany, zoology, entomology, chemistry, pathology, and ecology. You would study courses that deal with the environment, the threats to it, and the solutions. In addition, you will study governmental regulations and laws that deal with the environment. The Environmental Protection Agency (EPA) or the U.S. Department of Agriculture (USDA) are potential governmental employers for students with an interest in environmental sustainability.

Figure 16–15

Many biotechnology careers involve working with environmental issues.

Careers in Social Science

A field that is growing in importance is that of social science in biotechnology. Social science is the study of human society and how people interact with each other. This area may include fields in economics, education, and politics. As you have already studied, people have a tendency to be a little frightened of such technology as gene splicing, cloning, and the development of genetically modified organisms. Social scientists help in alleviating these fears by informing and educating the public on issues concerning biotechnology (Figure 16–16).

Most of the jobs in this area would require at least a 4-year degree. Course work will be concentrated in economics, communications, education, leadership, and sociology. A major component will be the acquiring of computer skills that will allow you to analyze data and communicate with people from all across the world. Several government agencies deal with agriculture, and many advisers are needed to help devise biotechnology policy for the agricultural sector of the economy.

Careers in Political Science and Law

Students with an interest in the development of biotechnology and how it impacts society may also consider careers in politics and law. In addition to explaining biotechnology

Figure 16–16
Informing the public is a crucial job in biotechnology.

© Photodisc/Getty Images

advances to the general public, there is a need for scientists who consider the legal and regulatory implications of emerging life technologies. After earning a bachelor's degree in a biotechnology-related field, some students continue on to law school or gain advanced degrees in political science.

Lawyers with a background in science may help to sort out questions of intellectual property (IP) and technology transfer related to biotechnology. In addition to helping inventors protect their investment in new technologies by drafting patents, IP specialists develop licenses and other agreements to facilitate sharing of related technologies. Some IP experts are interested in developing new methods of sharing discoveries that have the potential to benefit all of humanity. Political scientists with a background in biotechnology can help government officials draft appropriate legislation and implement science-based regulatory policies. Regulations that are either too strict or too lenient prevent technologies from helping humanity. Without legal and policy experts, none of the advances made in research and development will ever make it to the marketplace or benefit society.

The number of people who grow up on a farm has diminished at a steady rate for many years, and this phenomenon will continue. This means that there is a large and growing segment of our society that knows little about the agricultural industry. The field of agricultural education is an important segment of that industry (Figure 16–17). The mission of agricultural education is to teach people about agriculture, and this certainly includes biotechnology. People who communicate about agriculture through the media, those who operate the Cooperative Extension Services, university professors, and middle and high school agriculture teachers are all involved in agricultural education.

The shortage of middle and high school agriculture teachers is rapidly becoming a serious problem. Some states have difficulty each year in finding qualified people to fill these positions. If you enjoy working in all the areas of agriculture and you enjoy working with people, you may be suited to becoming a middle or high school Agriculture Education teacher. Your agriculture teacher is a good source of information on how to become a teacher in your state.

Courses in agricultural education that emphasize agriscience and biotechnology will be of great benefit to you. Here you will study the science behind the production

Figure 16–17
There is a great critical shortage of agricultural education teachers.

of plant and animals as well as the fascinating field of bio-technology. You will get the opportunity to experience science in action as a result of your **Supervised Agricultural Experience Program (SAEP)**. You will also have the opportunity to test your skills in National FFA Organization (FFA) programs.

Many community colleges have programs in agriscience and biotechnology (Figure 16–18). As a matter of fact, one of the fastest-growing fields of study at 2-year colleges is that of laboratory technician. At these schools, you will study chemistry, biology, zoology, and botany along with other related courses. In addition, you will study advanced methods of laboratory techniques and greenhouse management. The program of study will also include courses in the use of computers.

For those who attend a 4-year college or university, there are more technical careers available. At the university, among the areas you will study are chemistry, biology, botany, entomology, agricultural economics, and plant, animal, and soil science. Most universities have student organizations that sponsor activities centered around your area of study. These may include the Agronomy Club, the Horticulture Club, Collegiate FFA, or the Collegiate 4-H.

If you have the ability and would like to continue your study of biotechnology, then you might consider a graduate degree in the sciences. With a PhD degree, you will be able to work with the world's most knowledgeable people in the area of biotechnology. Both master's and doctoral graduate students

Figure 16–18
Most careers in biotechnology will require education beyond high school.

take courses in advanced areas of biotechnology, specializing in genetics, biochemistry, entomology, microbiology, pathology, and or any number of related life science or engineering topics. Graduate students conduct independent research projects and publish their results in academic journals and by writing formal research reports in order to receive their degree (master's thesis or PhD dissertation). Graduate programs require students to focus on a specific research challenge and to develop a deep understanding of the chosen topic area.

Summary

Making a career path choice is a big decision and requires a lot of thought and research. If you are interested in addressing challenges in health care, agriculture, or environmental sustainability, you may want to consider choosing a biotechnology career path. The field of modern biotechnology offers a wide variety of interesting, well-paying jobs, with academic preparation that has application in other science-related fields, such as the health professions. Students in biotechnology have many options, including careers in research, teaching, community outreach, government, and law. Biotechnology is tied to the future of medicine, agriculture, and the management of natural resources, and job growth is expected to continue in related fields. Keep reading about current events and developments in biotechnology, and it just may be that you will find a job that is perfect for you.

CHAPTER REVIEW

Student Learning Activities

1. Make an appointment to visit your guidance counselor to discuss your abilities and aptitudes. List all the things you enjoy doing and try to envision a career that will allow you to use your interests, abilities, and aptitudes. Decide on several possible careers and obtain more information about the careers that interest you.

2. Conduct an Internet search for job openings in the biotech industry. Write a report about the qualifications for the job, the duties of the job, and your opinion as to whether you think this would make a good career. Be thorough with your answers and report to the class.

3. Spend the day with a person who works in a career you might like. Make a list of all the things you liked and disliked about the job. Share your experiences with the class.

4. List all the activities in FFA that would help you prepare for a career. Which of these activities deal in some way with biotechnology?

5. Talk with your agriculture teacher about a career as a teacher. Ask him or her to explain the benefits and disadvantages of the career.

Fill in the Blanks

1. The _____ rewards of any job are usually _____ tied to the amount of _____ it takes to secure and _____ in that job.

2. Major advancements in the biotechnology industry have been made in such areas as _____ development, human and animal _____, agricultural _____, and _____ protection.

3. Many employees in the biotechnology industry start as _____ _____ with an associate's degree and work toward their _____ degree while _____.

4. Plant _____ usually involves working with a _____ scientist who has a goal of producing a particular new _____ _____ that may be developed into a new _____.

5. A media prep technician operates _____, pH _____, stir plates, _____, filtration equipment, open _____ tanks, and _____.

6. Once the research has been _____ and approval has been _____ to use a new biotechnology, it is up to the _____ industry to make a product commercially _____.

7. _____ of pharmaceuticals is a tightly _____ process that must follow _____ _____ _____ and all Food and Drug Administration (FDA) _____.

8. Companies are always in need of _____ employees to _____ new products, potential _____, and market _____.

9. All aspects of _____, particularly those that use new _____, must be _____ to ensure that the _____ is properly protected.

10. Social science is the study of _____ _____ and how people _____ with each other, and this area may include fields in _____, _____, and _____.

True or False ....

1. In any career path, the ability to problem solve, follow written instructions, communicate clearly, and work well with other people is critical.

2. The career opportunities in biotechnology are limited.

3. Europe leads the world in the research, development, and commercial use of biotechnology products.

4. You must have an advanced degree, such as a master's degree or a PhD, in order to have a successful career in biotechnology.

5. Most research scientists must work closely with other researchers in a team setting.

6. An engineer usually plans how the product will be manufactured or produced.

7. A marketing specialist determines how much of the product or technology can be sold and at what price.

8. A customer service representative may teach the proper use of the product as well as help in the maintenance of the item.

9. Lawyers with a background in science work to sort out questions of intellectual property and technology transfer related to biotechnology.

10. There are few jobs available in the field of agricultural education.

Discussion ....

1. List some questions you should ask yourself when thinking about what kind of career you want to pursue.

2. Describe the education and training needed to become a research scientist.

3. List some entry-level jobs in biotechnology.

4. Explain why all members of a research team, including those employees holding entry-level jobs, are valuable.

5. Why do biotechnology companies hire technical writers?

6. What skills are necessary for someone pursuing a career as a biostatistician?

7. Why are careers in natural resources and the environment going to become more important in the future?

8. Explain why agricultural education is an important segment of the agriculture industry.

9. What does SAEP stand for? What are the benefits of having a SAEP?

10. Which career in the biotechnology field is most interesting to you? Why?

Glossary

1000 Genomes Project \ A project of the governments of the United States, the United Kingdom, China, and Germany that is working to expand knowledge of human genomic variation.

Adult stem cells (ASCs) \ Cells from an adult organism that can divide into different types of cells.

Agronomic traits \ Plant traits that make them useful to humans.

Alleles \ A pair of genes that controls a specific characteristic inherited from each parent.

Amino acid \ Organic substance from which organisms build proteins or the end product of protein decomposition.

Anaphase \ The third phase of mitosis during which the pairs of chromatids separate into an equal number of chromosomes and the centromeres duplicate.

Anemia \ A deficiency of hemoglobin, iron, or red blood cells.

Animal and Plant Health Inspection Service (APHIS) \ A federal agency that governs the testing of bioengineered plants. It also monitors the health of plants and animals grown or brought into the country.

Antibiotic \ Germ-killing substances produced by a bacterium or mold.

Antibodies \ The very specific biological substances that the body manufactures to combat specific diseases following an attack of a disease or a vaccination.

Antigens \ Greatly weakened disease-causing organisms.

Applied research \ The process of taking the discoveries learned in basic research and finding ways to make use of the knowledge.

Artificial insemination \ The deposition of spermatozoa in the female genitalia by artificial rather than natural means.

Artificial vagina \ A rigid tube that is used to collect semen for artificial insemination.

Asexual reproduction \ A type of cloning in the plant world that more likely ensures that a desirable species will be reproduced.

Associate's degree \ A degree awarded to a person completing a prescribed program of study that usually consists of 2 years of study above the high school level.

Assortment \ The varitty of traits passed on from one generation to another.

Autotransplantation \ The transplanting of cells or tissues that originated in the body of the organism that receives the transplant.

Auxins \ Plant hormones.

Bachelor's degree \ A degree awarded to a person completing a prescribed course of study that usually consists of 4 years of study above the high school level.

***Bacillus thuringiensis* (Bt)** \ A soilborne bacteria that effectively kills insects.

Basic research \ The process of investigating and understanding how nature functions.

Beta carotene \ A substance that provides the yellow or orange color to fruits and vegetables and is a main source of vitamin A.

Biodiesel \ A nonpolluting, biodegradable liquid fuel that is obtained from renewable raw materials and can be used to replace fossil diesel fuel.

Biodiversity \ The number and variety of organisms found within a specified geographic region.

Biofortified \ Food that has had additional nutrients added by using biotechnology.

Bioinformatics \ The use of information technology to store, analyze, sort, label, and share the many DNA sequences generated by genome sequencing projects.

Biolistics \ A technique that blasts DNA into the cells using tiny bullets composed of tungsten or gold particles with DNA attached.

Biomanufacturing \ Manufacturing through the use of biotechnology, such as gene splicing.

Biomass \ Material from organisms that can be converted to energy.

Biopiracy \ Illegal commercial development of naturally occurring biological material, such as plant substances or genetic cell lines.

Bioprospecting \ The legal collection of biological samples that can help medicinal and scientific research.

Bioremediation \ An environmental improvement process whereby living organisms can be used to consume and convert pollutants to harmless substances.

Biosensors \ Devices that combine biological and electronic systems to detect or measure small amounts of specific substances, including environmental pollutants.

Biostimulation \ Adding nutrients such as nitrogen and phosphorus to stimulate the growth of naturally occurring microbes.

Biotech \ The use of technology to influence biological functions.

Biotechnology \ (1) The manipulation of living organisms or parts of organisms to make products useful to humans. (2) Using knowledge of cells to modify their activities in order to make living organisms more effective in serving people. (3) Deals with the manipulation of the genes of organisms to alter their behavior, characteristics, or value.

Blastula \ An embryonic stage where the cells are dividing but have not begun to differentiate.

Blastula stage \ The period of growth in embryo development where the cells are dividing but have not begun to differentiate.

Bovine somatotropin (BST) \ A hormone composed of protein produced in the pituitary gland of cattle which helps control the production of milk.

Bulbs \ The subterranean buds of some plants, which has a short stem bearing overlapping, membranelike leaf bases, as in onions and tulips.

Callus cells \ Cells that have not been genetically programmed to become a certain type of plant tissue (undifferentiated).

Cambium \ The actively growing cells between the bark and the wood in a tree or shrub.

Canning \ Method of food preservation that heats food in cans or jars to temperatures high enough to destroy the microorganisms in the food, drive out the air inside the container, and seal the container so that no other microorganisms can get in.

Carbohydrates \ Any of certain organic chemical compounds of carbon, hydrogen, and oxygen that are used to create energy.

Career \ A summation of all the jobs a person may hold in his or her life.

Cell \ The ultimate functional unit of an organic structure, plant, or animal. It consists of a microscopic mass of protoplasm that includes a nucleus surrounded by a membrane. In most plants, it is surrounded by a cell wall.

Cell differentiation \ The changing of cells into different types of tissues.

Cell membrane \ Structure composed of proteins and lipids and that contains the cytoplasm and the nucleus.

Cell wall \ A structure, found only in plant cells, composed of cellulose and providing some support for the entire plant.

Cellulosic ethanol \ Ethanol manufactured from cellulose.

Chloroplasts \ Plastids that use the energy of the sun to make carbohydrates and contain chlorophyll that gives plants their green color.

Chromoplasts \ Plastids that manufacture pigments that give fruits their color and also give leaves their brilliant color in the fall.

Chromosomes \ A microscopic, dark-staining body, visible in the nucleus of the cell at the time of nuclear division, that carries the genes, arranged in linear order. Its number in any species is usually constant, and it serves as the bridge of inheritance, that is, the sole connecting link between two succeeding generations.

Chymosin \ The most important enzyme in rennin.

Climatic tolerance \ Ability of crops to withstand the weather extremes in a given area.

Cloning/clones \ Production of a genetically identical offspring by artificial means.

Codominant \ Genes are of equal power, and neither gene is dominant.

Codon \ A nucleotide that tells the cell to build a certain type of amino acid.

Community college \ A relatively small college that usually offers 2-year plans of study.

Competent cells \ Cells that have been treated to make them able to receive vector DNA.

Conception \ The beginning of gestation; when a sperm fertilizes an ovum.

Confidence level \ The point at which research results are consistently proven accurate and accepted as reliable.

Control group \ One of two observation groups that will be identical to the other group with the exception of the research treatment.

Controller genes \ Gene sequences that regulate the gene's activity.

Controller sequence \ Areas that actually turn on or off a particular gene.

Corms \ Enlarged fleshy bases of a stem, bulblike but solid, in which food accumulates.

Corpus luteum \ Active tissue that develops on the ovary and produces progesterone if conception occurs.

Correlation \ The relationship of one set of data to another set of data.

Curds \ Solids separated out of milk.

Cytometer cell sorter \ An instrument used to sort female (X) sperm from male (Y) sperm according to the amount of light given off by the chromosome sperm as the female (X) chromosome gives off more light.

Dam \ The female parent.

Data \ The bits of information obtained and recorded that will form the basis for accepting or rejecting the hypothesis.

Deoxyribonucleic acid (DNA) \ *See* DNA.

Developing countries \ Countries that are in the process of developing in terms of infrastructure, food production, and so on.

Dietary deficiencies \ Lack of one or more of the essential nutrients in a human or animal's diet.

Differentiation \ The point at which stem cells change and develop into specific types of cells.

Diffusion \ A process by which molecules in solution pass through the cell membrane from a region of a higher concentration of molecules to a region of lower concentration of molecules.

Division \ A method of obtaining new plants by cutting the plant part into sections and growing a new plant from each section.

DNA (deoxyribonucleic acid) \ A genetic proteinlike nucleic acid on plant and animal genes and chromosomes that controls inheritance.

DNA fingerprinting \ Analyzing DNA sequences that are identical in any tissue in the body. This process is used for matching potential organ donors, identifying relatives, identifying crime victims, and helping establish guilt in criminal cases.

DNA probe \ A radioactive instrument that adheres to complementary sequences on certain

DNA fragments that, when exposed to photographic film, reveal an image of the DNA.

Donor cows \ A cow that is of unusual value as a breeding animal. The cow is used to obtain embryos that are transferred into a cow of lesser value.

Double helix \ Description of the two spiral-shaped strands of phosphoric acid and deoxyribose found within the nucleus of cells.

Drying \ Process of food preservation that removes moisture in the food necessary for bacterial growth.

Ecology \ A branch of science concerned with the interrelationships of organisms and their environments.

Edible plant vaccines \ Edible plant foods that have been genetically altered to contain antigens that will create a permanent or semipermanent resistance to disease.

Ejaculation \ The discharge of semen from the reproductive tract of the male.

Electron microscope \ A type of microscope that uses beams of electrons to illuminate the object.

Electroporation \ A process whereby an electric charge causes the cell membrane to become permeable to the DNA.

Electrophoresis \ A method of sorting DNA fragments according to size by subjecting the fragments to a direct electric current that draws the DNA through a gel. The negatively charged DNA is sorted because the smaller pieces of DNA move more rapidly through the gel.

Embryo transfer \ The process of removing an embryo from a superior female and implanting it into an inferior female.

Embryonic stem cells \ Cells from an embryo that are capable of producing any of the cells in an animal's body.

Endangered species \ Animal species that are in danger of becoming extinct.

Electrophoresis \ A method of sorting DNA fragments according to size by subjecting the fragments to a direct electric current that draws the DNA through a gel. The negatively charged DNA is sorted because the smaller pieces of DNA move more rapidly through the gel.

Endoplasmic reticulum (ER) \ A large webbing or network of double membranes that are positioned throughout the cell whose function is to provide the means for transporting material throughout the cell.

Enhanced bioremediation \ The group of techniques in which nutrients, microorganisms, or other materials are introduced to a contamination site to accelerate the cleanup process.

Enucleated oocyte \ An unfertilized ovum that has had its nucleus, including the DNA, removed.

Environment \ The conditions and circumstances surrounding an organism.

Enzymes \ A large, complex protein molecule produced by the body that stimulates or speeds up various chemical reactions without being used up itself; an organic catalyst.

Epigenetic inheritance \ An observed change in phenotype or underlying gene expression without a corresponding change in the DNA sequence.

Estrus \ The period that a female animal is ready for mating.

Estrous cycle \ The reproductive cycle in nonprimates.

Ethical dilemmas \ A problem dealing with ethics that seems to have a difficult solution.

Ethics \ The principles of right conduct that govern the conduct of a person or the members of a profession.

Eukaryotic cells \ A type of cell that contains genetic material within the confines of a membrane-enclosed nucleus; all plants and animals are made up of eukaryotic cells.

Exons \ Codon sequences that transcribe amino acids.

Experiment stations \ Research centers operated by land grant colleges to enable producers access to the latest knowledge on animal and crop production.

Experimental design \ A way to test a hypothesis in which a treatment group is compared to a control group.

Explant \ Plant tissue, containing the meristem cells, that is taken from the very end of a stem or root.

Extenders \ A substance, such as milk, egg yolk, glycerine, and/or antibiotics, that is added to semen to dilute it as well as to protect the sperm during freezing and provide nourishment for the sperm.

Fact \ A phenomenon that has been proven beyond a doubt.

Fermentation \ The processing of food by means of yeasts, molds, or bacteria.

Flow cytometer \ A device that separates the sperm into different tubes according to the amount of light given off.

Follicle \ A small blisterlike development on the surface of the ovary that contains the developing ovum.

Follicle-stimulating hormone (FSH) \ A hormone used to induce estrus.

Food allergy \ An adverse reaction such as rashes or sickness caused by a food component.

Food and Drug Administration (FDA) \ A federal agency responsible for monitoring the safety of food and drug products.

Food chain \ The progression of animals eating plants and animals eating the animals that eat the plants.

Forward genetics \ A process in which scientists observe the physical characteristics to find organisms with interesting or useful characteristics.

Freeze-drying \ Process of food preservation by quickly freezing the food and then placing it in an air-tight chamber where a vacuum is created and all moisture is removed.

Gametes \ Sex cells that are formed in a process known as meiosis.

Genes \ The simplest unit of inheritance. Physically, each gene is a nucleic acid with a unique structure. It influences certain traits.

Gene splicing \ The process of removing and/or inserting genetic material in order to change an organism's trait(s).

Gene therapy \ (1) A medical procedure to treat and cure genetic disorders that includes modifying or replacing disease-causing genes. (2) Procedure to combat genetic diseases that involves the insertion of genes into the human genome that corrects a problem with a person's original genes.

Genetic code \ The order in which four chemical constituents are arranged in huge molecules of DNA; these molecules transmit genetic information to the cells by synthesizing ribonucleic acid in a corresponding order.

Genetic diseases \ Diseases that are caused by a problem in a person's genetic makeup.

Genetic engineering \ *See* Gene splicing.

Genetic improvement \ Genetically changing organisms to give them more highly desired characteristics.

Genetic Information Nondiscrimination Act of 2008 \ A law passed to prevent discrimination against individuals in regard to health insurance or employment, based on their genetic information.

Genetic makeup \ The genetic constitution, expressed and latent, of an organism.

Genetic mapping \ The genetic makeup of an organism that includes the specific location of genes on chromosomes.

Genetic sequencing \ Order in which genes are situated on a particular chromosome.

Genetic transfer \ A process of removing a desirable gene from one organism and transplanting it into another organism.

Genetically altered clones \ Organism that has been produced by cloning that also included the addition of special genes to develop specific desirable traits.

Genetically altered crops \ Plants that have had genetic material added or removed in order to improve such traits as drought resistance, insect resistance, disease resistance, and climatic tolerance.

Genetically engineered \ The process of removing genes from one organism and transferring them to another organism.

Genetically manipulated organism \ An organism that has had a gene artificially inserted or removed.

Genetically modified \ Organisms that have been genetically changed through the intervention of humans. This includes the process of selective breeding.

Genetically modified organisms (GMO) \ Those organisms that have had genetic material removed and/or inserted in order to change one or more particular traits of the organism.

Genetics \ (1) The science that deals with the laws and processes of inheritance in plants and animals. (2) The study of the ancestry of some special organism or variety of plant or animal.

Genome \ A complete set of chromosomes inherited as a unit from one parent.

Genomewide association studies (GWAS) \ A project where researchers have been working to identify specific genes and nearby genome variations that correlate with medically important traits, such as susceptibility to a genetic disease or the ability to metabolize a certain type of drug.

Genotype \ The actual genetic makeup of an animal.

Glyphosate \ Active ingredient in the nonselective herbicide called Roundup.

Golden Rice \ Genetically altered rice containing a beta carotene–generating gene from a daffodil plant.

Golgi apparatus \ An organelle, shaped like a group of flat sacs bundled together, whose function is to remove water from the proteins and prepare them for export from the cell.

Gonadotropin-releasing hormone (GnRH) \ A hormone that causes a female to come into estrus.

Good manufacturing practices \ Controls that are required to be followed in manufacturing pharmaceuticals.

Graduate school \ A component of a college or university that offers programs of study above the bachelor's degree.

Grafting \ The process by which plant material from two separate plants are joined together to form one plant.

Green Revolution \ A term referring to the tremendous advances in food production that began in the 1970s.

Halophytes \ Plants that do well in salty soil.

Hatch Act \ A federal law passed in 1887 that established agricultural experiment stations operated by the land grant colleges in every state.

Helix \ A strand of nucleotides of DNA arranged together, resembling a long twisted ladder.

Herbicides \ Substance developed to kill weeds.

Herbicide-resistant crop \ Genetically engineered crop that is tolerant of herbicides.

Heredity \ The traits that are passed to an offspring from the parents.

Heterozygous \ Both the genes in the allele are different.

High-yield farming \ Producing more per acre.

Homeostasis \ The ability of an organism's cell to remain stable when conditions around it are changing.

Homologous recombination \ A process based on the fact that a gene for a particular characteristic has a specific place or position on the DNA strand on the chromosome.

Homozygous \ Both the genes in the allele are the same.

Horizontal gene transfer \ A process in which microbes are natural "genetic engineers" transferring genetic material between cells.

Hormones \ A chemical substance produced by the body or introduced into the body to produce a specific effect.

Human Genome Project \ A project with the goal of completely mapping all the genes of the human body.

Human insulin \ A protein hormone secreted by the islets of Langerhans, in the pancreas, that helps the human body use sugar and other carbohydrates.

Hypothesis \ A belief based on a study of relevant scientific literature.

Law of independent Assortment \ A principle or law states that factors (genes) for certain characteristics are passed from parent to the next generation separate from other alleles that transmit other traits.

Immunoassays \ Tests that use antibodies from animal immune systems to detect specific pollution compounds.

Imprinting \ An observed change in phenotype or underlying gene expression, without a corresponding change in the DNA sequence.

In vitro fertilization \ The process of fertilizing the egg in the laboratory and outside the female body.

Indicator species \ Uses plants, animals, and microbes to warn us about pollutants in the environment.

Induced pluripotent stem cells (iPSCs) \ Stem cells that are derived from adult, differentiated cells through treatment with special proteins or viruses.

Inoculation \ Introduction into healthy plant or animal tissue of microorganisms to produce a mild form of the disease, followed by immunity.

Integrated farm management \ A new method of farm management that aims to reduce applications of chemicals by optimizing the combination and timing of all farm management activities.

Intellectual property (IP) \ Ideas, concepts, and inventions that individual scholars or groups of scholars create.

International HapMap Project \ An effort to characterize genomic similarities and differences between various groups of people.

Interphase \ A phase of cell division where the cell is at the correct size, the chromosomes are duplicated, and the cell is ready to divide.

Intron sequences \ Areas on the chromosomes that contain no codes.

Knockout animals \ Transgenic animals that have had an original gene replaced or "knocked out" of action by a new gene that does not function.

Law of independent assortment \ A law which states that factors (genes) for certain characteristics are passed from parent to the next generation and are separate from the other factors or genes that transmit other traits; this separation allows the tremendous amount of diversity among organisms.

Law of segregation \ A law, developed by Gregor Mendel in the nineteenth century, that says that the factors responsible for the traits from each parent are separated and then combined with factors from the other parent at fertilization.

Layering \ Propagating plants by covering a portion of the plant with soil or other material and encouraging the portion to root while still attached to the parent plant.

Legumes \ Plants that produce their own nitrogen through the use of bacteria in nodules on the roots.

Lesser-developed country \ A Third World country or a country with a population so great it will have difficulty producing enough food products to sustain its people.

Leucoplasts \ Plastids that provide storage for the cell and are abundant in seeds, providing nutrients for emerging plants and animals that eat the seed.

Luteinizing hormone (LH) \ A hormone injected into a female to make her come into estrus.

Lissome \ organelles that serve as a cleanup crew within the cell, breaking down proteins, carbohydrates, and other molecules as well as any foreign material, such as bacteria, that enter the cell.

Literature search \ One of the first steps in properly conducting research; it entails a thorough search for literary information on the research topic.

Lumen \ An inner space within a cell where ions, such as calcium, are stored and proteins are folded and modified.

Lysosomes \ Organelles that are the digestive units of the cell.

Markers \ Differences in the patterns on fragments of DNA.

Marketing \ All the activity involved with the selling and buying of a good or service.

Master's degree \ A degree awarded to a person who completes a prescribed program of study above the bachelor's degree.

Mechanical refrigeration \ Means of lowering temperatures in order to preserve food by slowing down or completely stopping the actions of any microorganisms in the food.

Meiosis \ The process whereby cells are divided into cells that contain only one-half of the chromosomes needed for the formation of the young animal.

Meristem \ Plant tissue in which the cells are still undifferentiated.

Mesenchymal stem cells (MSCs) \ Stem cells that can differentiate into a wide variety of cells.

Messenger ribonucleic acid (RNA) \ A substance within the nucleus of the parent cell that accepts a copy of the genetic code for the organism.

Messenger RNA (mRNA) \ A molecule of a type of RNA within the nucleus of the parent cell on which the code of genetic information is copied or transcribed in a procedure that is somewhat like the replication of DNA in mitosis.

Metaphase \ The second phase of mitosis during which the chromatids move toward the center of the spindle, where they connect themselves to the fibers of the spindle.

Microfilaments \ Organelles that are fine fiber-like structures composed of protein and help the cell to move by waving back and forth.

Microinjection \ A method of placing DNA into an embryo by means of a very tiny needle.

Micromanipulator \ A device that is mounted on a modified microscope and used to grasp the embryo while the genetic material is inserted using an extremely fine needle.

Microtubules \ Organelles that are small, thin hollow tubes, composed of protein, that give the cells their shape and assist the movement of chromosomes during cell division.

Minerals \ Chemical compounds or elements of inorganic origin that are used to build bones, blood, and other tissues as well as to help the immune system to function properly.

Mitochondria \ A peanut-shaped mitochondria that functions to break down food nutrients and supply the cell with energy.

Mitosis \ The process whereby growth and cell reproduction take place in newly created cells.

Morula \ Early embryonic cells that divide and group together to form a ball-shaped mass where the cells divide and clump into a mass in a process called cleavage.

Motility \ Active movement in artificial insemination of the sperm in a male's semen.

Mycotoxins \ Potent poisons produced by a fungus on grains or other feedstuffs.

National Bioethics Advisory Commission \ Gives advice to the president on matters concerning ethics in biotechnology.

Nitrogen bases \ Adenine (A), thymine (T), guanine (G), and cytosine (C), which are attached to each other at the center of the rung where the two halves of the DNA helix ladder are connected.

Nitrogen use efficiency (NUE) \ A measure of the rate a certain plant requires nitrogen.

Nonselective herbicides \ A knockdown herbicide that kills all plants it is applied to.

Nontunicate \ A type of bulb that has layers of outer scales that can be separated and planted to grow a new plant.

Nuclear transfer \ The process of removing the nucleus of an ovum and replacing it with the nucleus of a body cell.

Nucleoid region \ The region around the nucleus of a cell.

Nucleotides \ A unit of DNA made up of a sugar molecule, a phosphate molecule, and a nitrogen molecule containing chemicals called bases.

Oleophilic \ More easily attracted to oils than water.

Oleophilic bacteria \ Bacteria capable of breaking down both simple and complex hydrocarbons found in crude oil.

On-the-job training \ Training or education that is obtained while a person is employed. The training or education is related to succeeding or advancing in the person's job.

Oocyte \ The unfertilized egg or ovum.

Oogenesis \ The production of eggs in the female reproduction system.

Organ transplant \ Surgically removing an organ from a donor animal or human and inserting it into a recipient animal or human.

Organelles \ Small structures within the cytoplasm of cells with the specific purpose to support the cells corresponding to that of organs within an animal's body.

Origin of replication \ DNA material the can be replicated.

Osmosis \ A process by which water passes through the cell membrane from a region of high concentration to a region of low concentration; if the cell has relatively little water inside, water is drawn from outside into the cell through the cell membrane.

Patents \ Government issued guarantees that the person or persons inventing a device or process has sole rights to the invention or process.

Patents on life \ The act of receiving a government patent for the discovery or creation of a living organism.

Pathogens \ Disease-producing agents, such as viruses, bacteria, or protozoa.

Peer review \ A process where other scientists evaluate the research done by other scientists.

Personalized genomics \ The particular gene makeup of an individual.

Personalized medicine \ Medicine that is designed for a particular individual.

Pesticides \ A substance used to control insect, plant, or animal pests.

Pharmaceuticals \ Substances used to enhance the health of humans or animals.

Pharming \ A term combining the words *pharmaceutical* and *farming* that refers to growing organisms (usually genetically altered) for the purpose of producing pharmaceuticals.

PhD \ Doctor of philosophy; a degree awarded that is usually the highest educational degree available in a particular field of study.

Phenotype \ How genes are expressed or how an animal actually looks.

Phloem \ Inner bark; the principal tissue concerned with the translocation of elaborated food produced in the leaves, or other areas, downward in the branches, stem, and roots.

Phytochemicals \ Chemicals that require light to function.

Phytoremediation \ The process of plants or trees absorbing or immobilizing pollutants.

Plant breeding \ The selection and collection of seeds from desirable plants to encourage their traits in future plant production.

Plant-made pharmaceuticals \ Medicines that are derived from plant sources.

Plasmids \ Circular-shaped pieces of DNA that float freely in the cell's fluid.

Plastids \ Bodies in plant cells that contain photosynthetic pigments.

Pluripotent \ Cells that can divide into many different types of cells.

Polymorphisms \ The variation of traits that may exist between like organisms; for example, blood type.

Progeny testing \ The testing of the characteristics of offspring from an animal.

Prokaryotic cells \ A type of cell that contains genetic material that is not confined to a nucleus; the smallest of all cells and generally considered to be neither plant nor animal.

Promoters \ A controller gene sequence that may work in concert with hormones to start or stop an activity.

Pronucleus \ Either the nucleus of the sperm or the egg that contains half the chromosomes of the fertilized ovum.

Prophase \ The first phase of mitosis in which the chromatin appears, the nuclear membrane begins to dissolve, and the entire nucleus begins to disperse while a new structure called a spindle is formed. (In animal cells, the centrioles move to opposite sides of the cell.)

Prostaglandins \ Hormones that when injected into the recipient cow will make it come into estrus.

Protectant \ Substance such as glycerine that is added to semen before it is frozen.

Proteins \ Any of a large number of complex, organic compounds of amino acids that provide the building blocks needed for creating new cells and tissue.

Protoplast \ A unit of protoplasm in one cell.

Quality traits \ Traits that improve crop value to the consumer, whether the crops will be used to feed people or livestock.

Quarantine \ Process that requires animals to be kept in isolation for a period of time to make sure that they do not have a disease.

Quiescent \ A state of inactivity, such as that in a cell.

Quiescent cells \ Cells going through periods of inactivity.

Recipient cows \ A cow of ordinary value that will be implanted with the superior female's embryo.

Recombinant DNA \ DNA combined from different sources.

Refugia \ A process in which noninsect resistant plants are grown nearby, either mixed with the biotech crops or planted in large sections.

Reliability \ The ability to replicate a research study that has been done and come up with the same results.

Rennin \ A coagulant enzyme occurring particularly in the gastric juice of calves and also in some plants and lower animals.

Research \ The creation of new knowledge and usually involves gaining an understanding of natural phenomena.

Restriction analysis \ Comparison of DNA patterns.

Restriction endonucleases \ A group of enzymes that recognize specific DNA sequences and cut the DNA into pieces.

Restriction fragment length polymorphism (RFLP) \ A process whereby a scientist will cut a fragment of DNA from tissue to be compared using an enzyme that cuts DNA at a specific site.

Reverse genetics \ A process where scientists have some knowledge, perhaps gained from a genome sequencing (gene mapping) project, that leads them to investigate the biological function of a specific gene.

Rhizomes \ Elongated underground stems or branches of a plant that send off shoots above and roots below and are often tuber shaped.

Ribonucleic acid (RNA) \ The substance in the living cells of all organisms that carries genetic information needed to form protein in the cell.

Ribosomal RNA (rRNA) \ A specialized substance that synthesizes amino acids.

Ribosomes \ Organelles of very tiny structures that are the sites where protein molecules are assembled in the cell.

Rule of dominance \ If one gene is dominant, it will override the other.

Scientific method \ A systematic approach, generally adopted as the standard procedure for conducting research, that consists of seven deliberate steps: stating the problem, formulating a hypothesis, designing the research methodology, conducting the experiment, collecting and analyzing the data, drawing conclusions and implications, and writing the research report.

Scion \ In grafting, the upper part of a plant.

Selective breeding \ Reproducing only the superior plants.

Selective herbicides \ A herbicide that kills only certain types of plants.

Semen \ A fluid substance produced by the male reproductive system containing spermatozoa suspended in secretions of the accessory glands.

Separation \ A method of obtaining new plants by pulling apart a plant where it naturally separates for the production of new plants.

Significance \ The probability that a difference in research findings happened by sampling error or chance.

Single nucleotide polymorphisms (SNPs) \ Single base changes in the DNA that may vary between people.

Sire \ (1) The male parent. (2) To father or to beget; to become the sire of.

Sperm \ The male sex cell, produced by the testicles.

Sperm sexing \ Separating the female (X) cell from the male (Y) cell in order to produce only male or only female offspring.

Spermatogenesis \ The production of sperm in the male reproductive system.

Stacked traits \ Crops with more than one genetically engineered trait.

Statement of the problem \ The first step in the scientific method and should clearly provide a basis and rationale for conducting research.

Stem cells \ Cells that are programmed to begin the differentiation process of cell development.

Stolons \ Runners, or specialized stems, that grow on top of the ground and reach out horizontally.

Straws \ Small hollow tubes that contain frozen semen.

Stress tolerance \ The degree to which an organism is able to withstand stress.

Superovulation \ Process of injecting the donor animal with a follicle-stimulating hormone that will cause the animal to release several eggs instead of just one.

Supervised Agricultural Experience Program (SAEP) \ An after-school educational program connected with a high school Agricultural Education program that is aimed at gaining practical experience in some area of agriculture.

Surrogate mothers \ Female that becomes host mother but is not the biological mother.

Sustainability \ The capability of maintaining a particular operation indefinitely.

Synthetic biology \ The process where biological parts and entire living systems may be designed and synthesized in the laboratory.

Systemic pesticides \ Pesticides that are taken up by the root system and spread throughout the plant where they can get at the insects.

Technician \ A person who is a specialist at performing a particular technique.

Telophase \ The final phase of mitosis in which the chromosomes continue to migrate to opposite sides (poles) of the cell, the spindles disappear, new membranes surround the chromosomes, and two new nuclei are formed.

Theory \ An explanation of existing observations based on the conclusions of several different research studies.

Tissue culture \ A means of plant propagation that involves the use of a small amount of tissue from a plant to grow a new plant.

Totipotent cell \ A complete cell that has formed with all the genetic material necessary for its development into a complete organism.

Transcription \ The code of genetic information is copied to a molecule of mRNA in a procedure somewhat like the replication of DNA in mitosis.

Transducer \ An electronic instrument that measures physical change in the environment produced by a biosensor.

Transfer RNA (tRNA) \ A substance within the cytoplasm that translates the genetic information into the amino acid.

Transformed plants and animals \ Plants or animals that have been genetically modified.

Transgametic technology \ A method of genetic engineering that uses a specially treated viral vector before the membrane surrounding the egg has fully developed.

Transgenes \ Genes that are transferred from one organism to another.

Transgenic \ Organisms that have genetic materials inserted from a different species.

Transgenic organisms \ Organisms that have been developed by splicing genetic material from a different type organism.

Treatment group \ One of two observation groups that will be given a different treatment which the researcher intends to study.

T-test \ A tool used by scientists to analyze differences in samples; it involves complicated mathematical calculations that are generated by a computer resulting in a *t*-value output.

Tuber \ Thickened or swollen underground branch or stolon with numerous buds (eyes).

Tunicate \ A type of bulb that has dry outer layers of membranes that are the result of the previous year's growth and propagates naturally by growing small bublets around the bulb.

Vaccine \ A substance that contains live, modified, or dead organisms or their products that is injected into an animal in an attempt to protect the host from a disease caused by that particular organism.

Vacuoles \ Organelles that serve as storage compartments for the cell.

Validity \ A term used to describe whether a research study achieved what the researcher intended it to.

Vectors \ A mechanism for transporting the genetic material inside the host cell.

Vegetative propagation \ Increasing the number of plants by planting seed or by vegetative means from cuttings, division, grafting, or layering.

Vertical gene transfer \ The transfer of genetic information from parents to offspring.

Virions \ Viral particles that have some similarities to cells in that they contain genetic material and structural proteins.

Vitamin A \ Vitamin that is synthesized from beta carotene that helps to combat blindness.

Vitamins \ Organic substances that aid the immune system and assist in the creation of enzymes.

Water use efficiency (WUE) \ How well a plant efficiently uses water.

Whey \ Liquid separated out of milk.

Withdrawal periods \ Time from the application of the pesticide until harvest required to render the produce safe to eat.

Xenografts \ The graft of tissue from one species to another, such as using a pig valve in a human heart.

Xenotransplantation \ The transplantation of organs from a different species of animal.

Xylem \ The "plumbing" system that conducts water and dissolved mineral up the stems from the roots.

Yeast \ A yellowish substance composed of microscopic, unicellular fungi of the family *Saccharomycetaceae* that induces fermentation in juices, worts, doughs, and so forth.

Zygote \ The fertilized egg or ovum.

Index

Page references followed by 'f' indicate a figure
Page references followed by 't' indicate a table

CENGAGE
Learning™

Dear Instructor:

NO. 94936429

Thank you for your interest in Cengage Learning textbooks. We hope you find that the enclosed textbook meets your course needs. If, however, you decide NOT to adopt the book, we encourage you to do your students and the environment a favor by recycling it.

Group: 1466333

Recycling this instructor's copy is an easy way to help keep the cost of textbooks down and takes little time. Simply use the postage paid return label below to return the book to the publisher.

Thank you!

<u>To use this label to recycle this instructor copy, simply:</u>

1. Cut out the label below.
2. Tape to any standard mailing envelope.
3. Give it to your U.S. mail carrier or drop it off at the nearest U.S. Post Office retail window.

FROM:

_____

_____

_____

POSTAGE DUE COMPUTED BY
DELIVERY UNIT

TOTAL POSTAGE DUE $_____

MEDIA MAIL

MERCHANDISE RETURN LABEL
PERMIT NO. 500   INDEPENDENCE KY 41051
CENGAGE LEARNING   10650 TOEBBEN DRIVE

POSTAGE DUE UNIT
US POSTAL SERVICE
5106 MADISON PIKE
INDEPENDENCE KY 41051-9998

NO POSTAGE
NECESSARY
IF MAILED IN
THE UNITED
STATES